Tourism as a Pathway to Hope and Happiness

ASPECTS OF TOURISM

Series Editors: Chris Cooper (*Leeds Beckett University, UK*), **C. Michael Hall** (*University of Canterbury, New Zealand*) and **Dallen J. Timothy** (*Arizona State University, USA*)

Aspects of Tourism is an innovative, multifaceted series, which comprises authoritative reference handbooks on global tourism regions, research volumes, texts and monographs. It is designed to provide readers with the latest thinking on tourism worldwide and in so doing will push back the frontiers of tourism knowledge. The series also introduces a new generation of international tourism authors writing on leading edge topics.

The volumes are authoritative, readable and user-friendly, providing accessible sources for further research. Books in the series are commissioned to probe the relationship between tourism and cognate subject areas such as strategy, development, retailing, sport and environmental studies. The publisher and series editors welcome proposals from writers with projects on the above topics.

All books in this series are externally peer-reviewed.

Full details of all the books in this series and of all our other publications can be found on http://www.channelviewpublications.com, or by writing to Channel View Publications, St Nicholas House, 31–34 High Street, Bristol, BS1 2AW, UK.

ASPECTS OF TOURISM: 96

Tourism as a Pathway to Hope and Happiness

Edited by
Tej Vir Singh, Richard Butler and David A. Fennell

CHANNEL VIEW PUBLICATIONS
Bristol • Jackson

DOI https://doi.org/10.21832/SINGH8557
Library of Congress Cataloging in Publication Data
A catalog record for this book is available from the Library of Congress.
Names: Singh, Tej Vir, editor. | Butler, Richard, editor. | Fennell, David A., editor.
Title: Tourism as a Pathway to Hope and Happiness/Edited by Tej Vir
   Singh, Richard Butler and David A. Fennell.
Description: Jackson: Channel View Publications, [2023] | Series: Aspects
   of Tourism: 96 | Includes bibliographical references and index. |
   Summary: "This book explores the view that tourism can be a pathway to
   hope and happiness. It examines the role of tourism in preserving
   natural and architectural wonders, bringing out the best in tourists and
   locals and adding economic value if managed sustainably. It is a useful
   resource for students and researchers in tourism, psychology and
   philosophy"—Provided by publisher.
Identifiers: LCCN 2022033624 (print) | LCCN 2022033625 (ebook) | ISBN
   9781845418557 (hardback) | ISBN 9781845418540 (paperback) | ISBN
   9781845418564 (pdf) | ISBN 9781845418571 (epub)
Subjects: LCSH: Tourism—Psychological aspects. | Tourism—Philosophy. |
   Tourists—Mental health.
Classification: LCC G155.A1 T589146 2023 (print) | LCC G155.A1 (ebook) |
   DDC 910/.019—dc23/eng20221102
LC record available at https://lccn.loc.gov/2022033624
LC ebook record available at https://lccn.loc.gov/2022033625

British Library Cataloguing in Publication Data
A catalogue entry for this book is available from the British Library.

ISBN-13: 978-1-84541-855-7 (hbk)
ISBN-13: 978-1-84541-854-0 (pbk)

Channel View Publications
UK: St Nicholas House, 31–34 High Street, Bristol, BS1 2AW, UK.
USA: Ingram, Jackson, TN, USA.

Website: www.channelviewpublications.com
Twitter: Channel_View
Facebook: https://www.facebook.com/channelviewpublications
Blog: www.channelviewpublications.wordpress.com

Copyright © 2023 Tej Vir Singh, Richard Butler, David A. Fennell and the authors of individual chapters.

All rights reserved. No part of this work may be reproduced in any form or by any means without permission in writing from the publisher.

The policy of Multilingual Matters/Channel View Publications is to use papers that are natural, renewable and recyclable products, made from wood grown in sustainable forests. In the manufacturing process of our books, and to further support our policy, preference is given to printers that have FSC and PEFC Chain of Custody certification. The FSC and/or PEFC logos will appear on those books where full certification has been granted to the printer concerned.

Typeset by Nova Techset Private Limited, Bengaluru and Chennai, India.

This book is dedicated to the life, work and memory of Professor Tej Vir Singh

Contents

	Contributors	ix
	Preface *Tej Vir Singh*	xv
	Introduction *David A. Fennell*	xvii

Part 1 Theoretical and Philosophical Foundations

1	Moving from Positive Psychology to Positive Tourism: A Conceptual Approach *Metin Kozak*	3
2	The Search for Meaning in Life: The Role of Tourism *Carla Fraga and Vera Lúcia Bogéa Borges*	16
3	*Eat Pray Love* or *Total Recall*? Mindfulness and Tourism *Gianna Moscardo*	32
4	Volunteer Tourism: A Pathway to Hope and Happiness? *Reni Polus and Neil Carr*	49

Part 2 Destinations, Settings and Populations

5	Familiar Tourists as a Source for Hope, Happiness and the Good Life: *In Situ* Tourist Tales *David Bowen and Jackie Clarke*	65
6	Food Tourism through the Lens of Post-Materialism: Valuing the Cultural Ruralscape *Subhajit Das and Hiran Roy*	86
7	Family Travel, Positive Psychology and Well-Being *Mona Mirehie and Iryna Sharayevska*	104
8	The Trinidad Carnival and the Promotion of *Joie de Vivre* *Johnny Coomansingh*	119

Part 3 Adjustment and Change

9 Retreating towards Subjective Well-Being 135
   Melanie Kay Smith

10 Enrichment and Enlightenment from Engagement with
   History through Heritage-Based Tourism 152
   Michael Fagence

11 Navigating the New Normal: Restorative Tourism
   Experiences during Times of Crisis 170
   Sera Vada and Noel Scott

12 Tourism, Hope and Peace: A Counter-Discourse in Palestine 185
   Rami K. Isaac

13 Deep Peace and the Solo Wilderness Canoe Experience 202
   David A. Fennell

   Conclusion 227
   Richard Butler

   Index 236

Contributors

Vera Lúcia Bogéa Borges has been an associate professor in the Department of Tourism and Heritage at the Federal University of the State of Rio de Janeiro, Brazil, since 2013. She has a postdoctorate in Tourism (2021–2022) from the University of Girona, Catalonia, Spain. She holds a PhD in History from Rio de Janeiro State University, Brazil. She leads the Group for Interdisciplinary Research of Tourisms and Cities (INTERTUR). Her research interests are tourism and history. She is a member of the National Association for Research and Postgraduate Study of Tourism (ANPTUR) and of the National History Association (ANPUH).

David Bowen gained his first degree in geography from the University of Oxford, UK. His interest in academic study related to tourism started while working in Kenya. As a result, he studied for an MSc in Tourism at Surrey University, UK, and, after joining Oxford Brookes University, completed a PhD entitled 'Consumer Satisfaction on Long-Haul Tours'. David has developed and taught numerous university courses on tourism at undergraduate and postgraduate level. He is currently Reader and Head of Doctoral Programmes at the Oxford Brookes Business School. David's research interests focus on tourist consumer behaviour and tourism destination development. He has published on those and other topics in a range of academic journals and recently wrote the completely revised second edition of *Contemporary Tourist Behaviour: Yourself and Others as Tourists* (CABI, 2022).

Richard Butler is Emeritus Professor of Tourism in the School of Business at the University of Strathclyde, UK. He has published 25 books and nearly 100 journal articles and book chapters and was awarded the UN World Tourism Organization Ulysses Prize in 2016 for 'excellence in the creation and dissemination of knowledge'.

Neil Carr is a professor in the Department of Tourism at the University of Otago, New Zealand. His research focuses on issues of rights, welfare, and freedom within tourism and leisure experiences, with a particular emphasis on children and families, sex, and animals (especially dogs). The brains behind the wonky façade of Neil belong to Sarah and Ebony (and, before her, Gypsy and Snuffie).

Jackie Clarke is a Reader at Oxford Brookes University UK, she has an initial industry background in tour operation and aviation and is a Fellow of the Royal Geographical Society. Her research focuses on tourism that embraces tourist behaviour, marketing and intersections with geographical ways of thinking, most recently focusing on destinations and the phenomenon of place familiarity (with Bowen). Her research papers have been published across both tourism, consumer behaviour and marketing journals

Johnny Coomansingh was born in Sangre Grande, Trinidad, and started out as a high school teacher. After eight years of teaching mathematics and human and social biology, he went to work as an agricultural extension officer. Following his work in agricultural extension, he accepted a position in the Trinidad and Tobago Petroleum Company as a corporate communications officer. In 1996 he left Trinidad for the USA. In the USA he worked as a professor in the field of geography and tourism. He continues to write and engage in research activities in Trinidad and Tobago and the Caribbean region.

Subhajit Das is an assistant professor in the Department of Geography at Presidency University, India. He received his PhD in Geography from the University of Kalyani, India. His primary research interest is the geography of tourism, and specific areas are tourism and community development, community politics among tourism stakeholders, tourism walkability, and rural tourism. His research works have been published in edited books by international publishers such as Routledge and Channel View Publications. He has also published research articles in the journals of Elsevier and Taylor & Francis.

Michael Fagence is first and foremost a geographer. Drawing on the inspiration of Richard Hartshorne's ideas about 'areal differentiation' he has applied the processes of 'thinking geographically' to a number of fields, and in recent years his research has tended to focus attention on the telling of particular stories from the history of Australia through the lens of heritage-based tourism. His most recent appointment was as Honorary Research Fellow at the University of Queensland.

David A. Fennell is a professor in the Department of Geography and Tourism Studies at Brock University, Canada. He is the founding editor-in-chief of the *Journal of Ecotourism* and has published many articles and books, including *Sustainable Tourism: Principles, Contexts and Practices* (co-authored with Chris Cooper, 2020).

Carla Fraga has been an associate professor in the Department of Tourism at Juiz de Fora Federal University, Brazil, since 2021. She was a professor in the Department of Tourism and Heritage at the Federal University of the State of Rio de Janeiro, Brazil, from 2006 to 2021. She holds a DSc in Transport Engineering from the Transport Engineering Program of the Alberto Luiz Coimbra Institute of Postgraduate Engineering Study and Research of Rio de Janeiro Federal University, Brazil. Studying in a specialisation course in Neuroscience and Applied Psychology at Mackenzie Presbyterian University, Brazil, she leads the Transportation and Tourism Research Group. Her research interests are tourism, transportation, and neurosciences. She is a member of International Academy for the Development of Tourism Research in Brazil (ABRATUR), and of the National Association for Research and Postgraduate Study of Tourism (ANPTUR).

Rami K. Isaac, PhD, is a senior lecturer at the Academy for Tourism, Breda University of Applied Sciences, The Netherlands. In addition, he is an assistant professor at the Faculty of Tourism and Hotel Management at Bethlehem University, Palestine. Currently, he is the vice-president of the Research Committee 50 on International Tourism of the International Sociological Association. He has published numerous articles, book chapters, and edited volumes on tourism in conflict-ridden destinations, critical theory, and political aspects of tourism.

Metin Kozak holds a PhD degree in Tourism from Sheffield Hallam University, UK. He has contributed a wide range of articles to top-tier journals, conference papers in more than 40 countries, and over 30 books released by international publishers. He has been involved in several national and international research projects, particularly with his partners based in Europe, Asia, and the USA. He acts as the co-editor of *Anatolia* and is a member of the editorial/review boards for many international journals. His research interests are marketing and consumer behaviour in an interdisciplinary context. He is affiliated with Kadir Has University, Turkey.

Mona Mirehie is an assistant professor in the Department of Tourism, Event, and Sport Management at Indiana University–Purdue University, USA. Her research interests include tourism and well-being, recreational sport participation and well-being, and sociology of tourism with a focus on gender.

Gianna Moscardo is a professor of Tourism in the College of Business, Law, and Governance at James Cook University, Australia. Moscardo has qualifications in applied psychology and sociology and joined James Cook University in 2002. Her qualifications in applied psychology and sociology support her research interests in understanding how communities and organisations perceive, plan for, and manage tourism development opportunities, how tourists learn about and from their travel experiences, and how to design more sustainable tourism experiences. She has published extensively on tourism and related areas, with more than 250 refereed papers or book chapters. Moscardo has been invited to speak on issues related to tourism in New Zealand, South Africa, Botswana, Italy, Finland, Austria, Croatia, Wales, Spain and the USA. Her recent project areas include evaluating tourism as a tool for economic development in rural regions, tourist experience analysis, cross-cultural issues in tourism, designing effective tourist interpretation, tourist storytelling, and new approaches to tourism planning and management at the destination level.

Reni Polus is a PhD student in the Department of Tourism at the University of Otago, New Zealand. Her research interests are focused on spirituality within tourism and leisure experiences, particularly in volunteering, pilgrimage, dark tourism and heritage tourism.

Hiran Roy is a lecturer at the International School of Hospitality, Sports, and Tourism Management at Fairleigh Dickinson University, Canada. He holds a PhD in Management from the University of Canterbury, New Zealand. He has completed his MBA from the University of Guelph, Canada. Prior to joining academia, he worked extensively (for over 26 years) in leadership roles in the hospitality business industry on different continents. Hiran is also the recipient of the New Zealand Commonwealth Scholarship and Fellowship Plan, considered among the most prestigious awards in the world for his PhD program. His research interests include sustainable tourism, sustainable local food systems, local food marketing, city branding through food, hospitality luxury branding, corporate social responsibility in tourism and hospitality, and sustainability. Hiran's work has been published in a variety of book chapters, including books published by Routledge, Elsevier, Emerald, Springer Nature, Channel View Publications, and high-impact prestigious leading academic journals. He is an editorial board member of the *Journal of Global Scholars of Marketing Science*, *Journal of Foodservice Business Research*, *Journal of Human Resources in Hospitality & Tourism*, *Frontiers in Sustainable Tourism* and *Tourism and Hospitality* (MDPI).

Noel Scott (PhD) is Adjunct Professor of Tourism Management, in the Sustainable Research Centre, University of Sunshine Coast; Taylor's University, Malaysia, Mataram University, Indonesia and Edith Cowan University, Australia. His research interests include the study of tourism experiences, and destination management. He is a frequent speaker at academic and industry conferences. He has over 300 academic publications including 17 books. He has supervised 30 doctoral students to successful completion of their theses. He is on the Editorial Board of 10 journals, a member of the International Association of China Tourism Scholars and a Fellow of the Council for Australasian Tourism and Hospitality Education., professor at Clemson University, USA. Her research areas include technology-based leisure in contemporary families, leisure behaviour and well-being among individuals and families of diverse backgrounds, and use of technologies in recreation management (e.g. the new media/social network sites, gaming, cell phones, navigational devices).

Tej Vir Singh was the founding director of the Centre for Tourism Research and Development, Lucknow, India. He launched several journals and was the founding editor-in-chief of *Tourism Recreation Research*. He edited and co-edited 16 books and supervised 15 doctoral candidates who were awarded PhDs in Tourism.

Melanie Kay Smith (PhD) is an associate professor, researcher and consultant whose work focuses on urban planning, cultural tourism, wellness tourism, and the relationship between tourism and well-being. She is Programme Leader for BSc and MSc Tourism Management at Budapest Metropolitan University in Hungary. She has lectured in the UK, Hungary, Estonia, Germany, Austria and Switzerland, as well as being an invited keynote speaker in many countries worldwide. She was Chair of ATLAS (Association for Tourism and Leisure Education) for seven years and has undertaken consultancy work for the UN World Tourism Organization and ETC as well as regional and national projects on cultural and health tourism. She is the author of more than 100 publications.

Sera Vada is a Postdoctoral Research Fellow at the Griffith Institute for Tourism, Griffith University and has published in leading tourism and marketing academic journals. With a PhD in tourism management, her research and consulting interests include, positive psychology and tourist behaviour studies, destination marketing and diversification of tourist markets, tourism in Small Island Developing States (SIDS), tourism challenges in the Pacific and Chinese tourism. Sera also has over 10 years of industry experience across broad sectors of tourism, destination marketing, project management and tertiary education.

Preface

Tej Vir Singh

Tourism, Hope and Happiness: In Quest of a Good Life

This book deals with problems of 'how to live a good life'. All of us are interested in this question. Some would say 'wealth', others consider 'power and self'. There are people who would vote for 'honour' or 'fame'. The fact is that none of these afford happiness. You can be happy without them. A depressed millionaire may not be happy, while a poor, dancing girl on a faraway island might be. Happiness is not found, it is created.

We tested the truth of these statements in the field of action. Tourism, for all its popularity, is linked to hope and that has various shades of thought: desire is expectation, hankering for something, optimism. Hope is a feeling of strength. This strength is in the strength of the person's desire. Hope teams up with faith and believes in the impossible. The power of hope overcomes all obstacles. Hope is an optimistic state of mind.

Happiness is used in the context of mental or emotional states. It is also used in the case of life satisfaction and well-being. It includes positive emotion or intense joy. The contributions in this book act as illustrations of hope and striving for happiness; and provide clues for maintaining a good life.

Introduction
Reflection for Acceleration

David A. Fennell

Prologue

There are many journeys worth taking. For me, this has been one of them. I say that because this particular journey, investigating tourism as a pathway to hope and happiness, started with a seed planted by Professor T.V. Singh in November of 2020. While this is not a *bona fide* dedication section (see Singh, 2021), please allow a short moment of indulgence as a point of departure for what is to follow.

There are watershed moments in our personal and professional lives galvanised almost as if forged in steel. For a young scholar, mine came in correspondence with T.V. Singh in 1999. Out of the blue, T.V. (he never let me call him Tej) contacted me and offered kind words of encouragement – more like validation – for the type of research I was engaged in (ecotourism, tourism ethics, human nature, and later animal ethics). Implicit in his message was that although the world was being enveloped by the neoliberal audit culture – meaning adopting well-established themes of investigation would generate more citations and perhaps notoriety – there was a different way. 'Stay the course,' he argued, 'and we will make space for a different tourism world'. Professor Singh understood the road that our field was taking and saw it necessary to embrace a new platform based on ethics to broaden the horizon of possibilities, no doubt as a function of his own deeply held values and beliefs.

As our discussions on the book started to advance through the end of 2020 and the beginning of 2021, I could tell by his emails, broken and piecemeal, that his ailments were catching up with him. Nevertheless, despite his physical and cognitive challenges, he provided advice on authors, the title of the book, and the proposal, all while receiving, as he wrote, disapproval from his family because of his diminishing health. He tried to, in his own words, 'run the race', but was unable to in the end. After his passing, Professor Butler joined on as co-editor because he wanted to contribute to this project in honour of T.V. I am thankful to be riding shotgun alongside the two most influential figures in my development as a scholar.

The following metaphor from the Indigenous people (Haudenosaunee and Anishinaabe), on whose land I currently reside, provides an apt manner to define Professor Singh's impact and legacy:

> Over 1,000 years ago, the Five Nations were brought together in peace at Onondaga Lake by the Peacemaker and Hiawatha (Hayenhwátha'). Together they planted the Great Tree of Peace (Skaęhetsiʔkona) and created the Haudenosaunee Confederacy. This is where Skä·noñh began anew. The Tree of Peace is a metaphor for how peace can grow if it is nurtured. Like a tall tree, peace can provide protection and comfort. Like a pine tree, peace spreads its protective branches to create a place of peace where we can gather and renew ourselves. Like the White Pine, peace also creates large white roots (tsyoktehækęætaʔkona) that rise out of the ground so people can trace their journey to the source. (Indigenous Values Initiative, 2022)

I would like to think that T.V. would give a nod of approval to the use of yet another ontology (and metaphor) that explains tourism in a different light. In the same vein as the Tree of Peace, a large part of Professor Singh's legacy rests in how he was able to grow tourism through a nurturing hand, how he created a space for tourism to grow, and how he fuelled renewal through knowledge and innovation. His roots, and spirit, run deep through the pages which follow. T.V.'s son, Sagar Singh, wrote to me after his father's passing and said that it was his father's 'last wish to end on a note of hope and happiness'. His passing was a jolt to the field. Young scholars should know that we – all of us – stand on the shoulders of giants. T.V. was a giant who wielded tremendous influence over the pace and direction of tourism scholarship. We lost a friend, pioneer and sage. I cannot think of a better way to honour his many contributions to our field than through this book, which I believe would please his soul. He was very much the embodiment of the book's title, and it is to his memory that we press on. So, out of deep sadness, a pathway to hope and happiness emerges . . .

Introduction

The present volume is not the first foray into tourism as a force for inducing positive states and outcomes. Numerous studies can be found on subjective well-being (Kim *et al.*, 2015; McCabe & Johnson, 2013), positive psychology (Filep & Pearce, 2017; Nawijn, 2016), positive tourism (Mkono, 2019), mindfulness (Frauman & Norman, 2004; Stankov & Filimonau, 2021), as well as tourism and the good life (Pearce *et al.*, 2012). Studies also exist on the primacy of pleasure as a central component of the touristic experience (Fennell, 2009, 2018). Yet, even with these works in place, leading thinkers on the topic argue for an expanded agenda on tourism as a dynamic positive influence (Filep & Laing, 2019) in the face of such an overwhelming focus on the negative aspects of tourism. Such an emphasis is especially important as we grapple with important issues related to personal well-being, but also issues of greater scale, including the well-being of family, community and planet.

Given the recent focus on positive psychology, few scholars have investigated the concept of psychological capital – following from Bourdieu's (1977) seminal work on power, profit, advantage, and ranking within a field (through, for example, social, cultural, and physical capital) – as recognition of 'who you are' (Luthans et al., 2007). Studies that have been undertaken on psychological capital include a focus on management and organisational domains. Lin (2013) investigated employee job burnout among a sample of workers in international hotels. Tsaur et al. (2019) found that fun in the workplace has a positive effect on the psychological capital of tourism and hospitality workers. More recently, Mao et al. (2020) argue that in the face of concerns about health and job security due to COVID-19, psychological capital can be enhanced through the implementation of CSR strategies in all areas, including self-efficacy, hope, resilience and optimism. The same holds true for a newer incarnation of the capitals concept in the form of existential capital, which seeks to explore the inner dimensions of deeply personal, and often embodied, individual experiences (Nettleton, 2013). There is scope for a broader understanding of personal meaning in our touristic ventures, certainly beyond the basic visual and gustatory pleasures, perhaps harkening back to the search for the 'centre' as observed by Cohen (1979).

Additionally, few studies exist on the topic of hope and hopeful tourism. Ateljevic et al. (2007) discuss the creation of an Academy of Hope in Tourism, with these authors dominating the discourse on this concept. Pritchard et al. (2011: 949) define hopeful tourism as 'a values-led humanist approach based on partnership, reciprocity and ethics, which aims for cocreated learning, and which recognises the power of sacred and indigenous knowledge and passionate scholarship'. Tucker and Shelton (2018) discuss narratives of loss and hope that thread together as effective moods 'that matter' produced and performed in tourism destinations like Christchurch, New Zealand, after the considerable impacts of natural disasters. In such situations as those and also following the global disaster of the COVID-19 pandemic, the return and resumption of tourism will hopefully signal a new beginning or rebirth and act as a signal of hope about a more positive future.

The present volume reinforces the importance of positive psychology subjective well-being, positive tourism, mindfulness, and other optimistic themes and forces that help to define the tourist experience. As a point of departure, it interweaves these concepts into a discourse on hope and happiness whilst at the same time incorporating new themes into the theoretical mix. Central to the approach is recognition that philosophy and human nature should never be too far removed as mechanisms to more deeply explain and understand the human condition.

This book is divided into three parts: Part 1: 'Theoretical and Philosophical Foundations'; Part 2: 'Destinations, Settings and Populations'; and Part 3: 'Adjustment and Change'. Part 1 begins with a chapter by Metin Kozak, whose work on positive psychology flows into a discussion of positive tourism, which has origins in philosophical humanism (Filep *et al.*, 2017), as an area not well developed in tourism studies. Chapter 1 stands as a departure from the current literature on positive tourism in its creation of a more holistic approach to the concept, highlighting tourist behaviour and operators and communities. The rewarding nature of experience is explored through hedonia, eudaimonia, romantic relationships, humour, and flow, which are used to investigate how tourism can contribute to psychological and physiological benefits.

Chapter 2 by Carla Fraga and Vera Lúcia Bogéa Borges investigates tourism's role in the search for meaning in life. The authors accomplish this by urging us to more formally embrace neuroscience in better understanding tourists and tourism, how neuroscience and applied psychology can explain relationships that exist between hosts and guests (e.g. social ostracism, reciprocal altruism), and, finally, how motivation theory explains how tourism can be repositioned after the pandemic. Disciplinary and theoretical alliances need to be forged, Fraga and Borges argue, to open up new paradigms and possibilities to more fully understand the nature of tourism.

Mindfulness is investigated by Gianna Moscardo in Chapter 3 through an engaging encounter with two films, *Eat Pray Love* and *Total Recall*, which portray very different visions of tourism. Through definition and an exploration of different mindfulness types and how the concept is being used in tourism practice and research, Moscardo concludes that mindfulness has untapped potential in elevating the role of sustainability in theory and practice on many levels. However, a word of caution indicates that mindfulness is not synonymous with being ethical. On the contrary, the good comes by using a critical lens to consider the needs and interests of others and how our actions have damaging impacts on the natural world.

In Chapter 4 Reni Polus and Neil Carr sustain the relatively new discussion on volunteer tourism not as a form of altruism but rather as self-interest. Hope and happiness surface in the expression of an ethic of care, which in the future opens the door for tourists or volunteers to experience their own rewarding experiences based on the actions and behaviours of the tourists who came before them – *successive* reciprocal altruism as described by Fennell (2006). Reciprocal altruism emerges through multiple acts of trust and cooperation that lead to symbiotic relationships, suggesting that an interdisciplinary research agenda is needed to understand the complexity of human nature better.

Part 2 on destinations, settings and populations begins with Chapter 5 by David Bowen and Jackie Clarke, who embark upon an *in situ* investigation of the pathways to hope and happiness that tourists in Wales gain from engaging with familiar places. The theme of 'we just keep coming back' emerges as tourists find the unfamiliar in the familiar and the extraordinary in the ordinary in attractions like wild rivers. There are management implications in their findings that point to the belief that destinations can carve out a competitive niche that is not subject to fads or fashions. Familiar tourism shares characteristics with slow, creative and transformative forms of tourism.

In Chapter 6 Subhajit Das and Hiran Roy focus on how the foodscape in rural areas acts as a conduit to bring hope, happiness and other positive psychological states through a post-materialist lens. Theory is applied in the context of the Eastern Himalayan region of Sikkim on the Silk Route, where authentic homestay owners emphasise giving more space to tourists in heritage the ethnic lifestyle, traditions of food preparation, and cultural significance. Das and Roy argue that a 'geography of taste' represents a sense of place based on its foodscape in the relationship between local cuisine and cultural values.

There is a return to a discussion on positive psychology in Chapter 7 by Mona Mirehie and Iryna Sharayevska, not addressing it in the same vein as Kozak's survey of the theoretical foundation of positive psychology but rather in how it relates to family travel and well-being. Families, Mirehie and Sharayevska observe, often pursue travel to improve their relationships, escape from routine, create memories, improve communication, and continue family traditions. Their qualitative investigation of 18 full nesters (parents and their children who still live with them) from Indiana and South Carolina reveals two hedonic elements (positive emotions and relationships) and two eudaimonic elements (detachment and mastery). All four of these domains were found to be interrelated and co-dependent based on participants' accounts.

Chapter 8 by Johnny Coomansingh on the 200-year-old Trinidad Carnival and the promotion of *joie de vivre* is about a form of happiness based on 'gustatory excesses with raucous ribald revelry' that represents 'time outside of time'. Carnival is so thoroughly embedded in the Trinidadian psyche that life is something that happens as a diversion from this national celebration, and there is existential meaning that comes packaged with pleasure and happiness. It is the joy of just being alive, Coomansingh maintains, manifested in many different ways, including going against the rules of law, resistance, misbehaviour, sexual liberation and the release of pressure built up over the year. This chapter is a study on the emotional tie to a place where deep meaning is generated in the minds of tourists and locals through this yearly event.

Part 3 switches from a specific focus on destinations to adjustment and change in tourism destinations. Many of the chapters in this volume summon, albeit briefly, Ancient Greek philosophy to position their arguments on the pathway to hope and happiness in tourism. Melanie Kay Smith's account on retreating towards subjective well-being in Chapter 9 is a case in point, with Aristotle's concept of universal harmony acting as a philosophical anchor point that intersects both society and nature. Indeed, the focus on retreats that harmonise with the natural world emerges as a strong theme in Smith's discourse analysis of retreat websites, with personal or existential benefits reflecting 'rejuvenation', 're-energizing', 're-charging', 'replenishing' and 're-booting' in finding one's true self.

The book takes a radical turn in Chapter 10 based on Michael Fagence's depiction of the 'real' events surrounding the Australian bushranger-cum-outlaw Edward (Ned) Kelly and the gunfight that took place between him, three of his associates, and a police patrol in Stringybark Creek, Victoria, Australia. The utter darkness of this event, and the lore and legend that followed, are handled head-on by Fagence, who cleverly weaves in a positive psychology narrative through pilgrimage, Australian nationhood, and, finally, the celebration of duty on the part of the police force. What bubbles to the surface in this story is that tourism, if adequately choreographed, can act as a pathway to enrichment and enlightenment through changing attitudes, behaviours, and values.

Eudaimonic and well-being themes are sustained in Chapter 11 by Sera Vada and Noel Scott, which is based on the challenges presented by COVID-19 – challenges that have increased isolation and distanced us from the ones we love. Through the use of a tourist well-being framework, the authors reveal how engaging in domestic tourism, especially with valued natural attractions, has the potential to restore our psychological and physiological health and well-being. The study has direct implications for tourism marketers and managers in finding the best fit between supply and demand and achieving levels of repeat visitation.

Space and place take centre stage in Chapter 12 in Rami K. Isaac's portrayal of the antagonistic relationship between Palestine and Israel and how tourism can be a pathway to peace, hope and happiness in this region. In particular, alternative tourism is used as a counter-discourse through constructing narratives of hope. As such, tourism can help tell the story of an occupied Palestine through hopeful narratives while at the same time challenging conventional perspectives. The objective is to create spaces for enhanced engagement between Israelis and Palestinians to overcome the historical-cultural impediments in creating a new, shared trajectory.

Finally, Chapter 13 by David A. Fennell, based on a solo wilderness canoe trip to the wilds of Temagami, Ontario, Canada, views peace not from a macro-level perspective (i.e. the broad structural reasons why peace has disintegrated because of security and development issues), as most studies to date adopt, but rather from an atomistic perspective.

Fennell is after the psychological or existential meaning of deep peace as a multidimensional state of being based on one's immersion into an activity leading to a profound sense of inner harmony, wisdom and contentment. Deep peace can play a role as metatheoretical 'connective tissue' not only in linking to other tourism and outdoor recreation concepts and experiences (e.g. specialisation and serious leisure), but in broadening the canvas of what these touristic events mean to us and us alone.

This latter chapter, and many others in this volume, speak to the primacy of the individual in the act and outcomes of tourism even as tourism has implications on much grander scales. But isn't the atomistic (i.e. a focus on the individual) level *the* central place at which to unravel the mysteries of tourism? Isn't tourism a wonderful expression of our human nature, especially when it comes to the intricacies of pleasure, pain, meaning, as well as hope and happiness. There is power, and I would say promise, in the individual, despite the consistent suppression of the legitimacy of individualism and humanism, as we increasingly become commercial and conformist entities through the onslaught of a pervasive corporate ethos (Saul, 1995).

We therefore urge scholars to walk the pathway of hope and happiness, not only in the following pages, but also in accelerating knowledge in this much-needed area of enquiry. Indeed, in so doing, we fulfil Professor Singh's wishes of a much different, and perhaps better, tourism world down the road.

References

Ateljevic, I., Morgan, N. and Pritchard A. (2007) Editors' introduction: Promoting an academy of hope in tourism enquiry. In I. Ateljevic, A. Pritchard and N. Morgan (eds) *The Critical Turn in Tourism Studies: Innovative Research Methodologies* (pp. 1–8). Amsterdam: Elsevier.

Bourdieu, P. (1977) *Outline of a Theory of Practice*. Cambridge: Cambridge University Press.

Cohen, E. (1979) A phenomenology of tourist experiences. *Sociology* 13 (2), 179–201.

Fennell, D.A. (2006) Evolution in tourism: The theory of reciprocal altruism and tourist–host interactions. *Current Issues in Tourism* 9 (2), 105–124. https://doi.org/10.1080/13683500608668241.

Fennell, D.A. (2009) The nature of pleasure in pleasure travel. *Tourism Recreation Research* 34 (2), 123–134.

Fennell, D.A. (2018) On tourism, pleasure and the *summum bonum*. *Journal of Ecotourism* 17 (4), 383–400.

Filep, S. and Laing, J. (2019) Trends and directions in tourism and positive psychology. *Journal of Travel Research* 58 (3), 343–354.

Filep, S. and Pearce, P. (eds) (2017) *Tourist Experience and Fulfilment: Insights from Positive Psychology*. Abingdon: Routledge.

Filep, S., Laing, J. and Csikszentmihalyi, M. (eds) (2017) *Positive Tourism*. Abingdon: Routledge.

Frauman, E. and Norman, W.C. (2004) Mindfulness as a tool for managing visitors to tourism destinations. *Journal of Travel Research* 42, 381–389.

Indigenous Values Initiative (2022) The Great Tree of Peace (Skaęhetsi?kona). See https://indigenousvalues.org/haudenosaunee-values/great-tree-peace-skaehetsi%CB%80kona.

Kim, H., Lee, S., Uysal, M., Kim, J. and Ahn, K. (2015) Nature-based tourism: Motivation and subjective well-being. *Journal of Travel & Tourism Marketing* 32 (sup. 1), S76–S96. https://doi.org/10.1080/10548408.2014.997958.

Lin, T.-L. (2013) The relationships among perceived organization support, psychological capital and employee's job burnout in international tourist hotels. *Life Science Journal* 10 (3), 2104–2112.

Luthans, F., Youssef, C.M. and Avolio, B.J. (2007) *Psychological Capital: Developing the Human Competitive Edge*. New York: Oxford University Press.

Mao, Y., He, J., Morrison, A.M. and Coca-Stefaniak, J.A. (2020) Effects of tourism CSR on employee psychological capital in the COVID-19 crisis: From the perspective of conservation of resources theory. *Current Issues in Tourism* 24, 2716–2734. https://doi.org/10.1080/13683500.2020.1770706.

McCabe, S. and Johnson, S. (2013) The happiness factor in tourism: Subjective wellbeing and social tourism. *Annals of Tourism Research* 41, 42–65.

Mkono, M. (2019) *Positive Tourism in Africa*. Abingdon: Routledge.

Nawijn, J. (2016) Positive psychology: A critique. *Annals of Tourism Research* 56, 151–153.

Nettleton, S. (2013) Cementing relations within a sporting field: Fell running in the English Lake District and the acquisition of existential capital. *Cultural Sociology* 7 (2), 196–210.

Pearce, P., Filep, S. and Ross, G. (2012) *Tourists, Tourism and the Good Life*. Abingdon: Routledge.

Pritchard, A., Morgan, N. and Ateljevic, I. (2011) Hopeful tourism: A new transformative perspective. *Annals of Tourism Research* 38 (3), 941–963.

Saul, J.R. (1995) *The Unconscious Civilization*. Toronto, ON: Anansi.

Singh, S. (2021) Shradhanjali: Remembering TEJ VIR SINGH (June 1929 – April 2021). *Tourism Recreation Research* 46 (2), 143. https://doi.org/10.1080/02508281.2021.1926094.

Stankov, U. and Filimonau, V. (2021) Here and now – the role of mindfulness in post-pandemic tourism. *Tourism Geographies*. https://doi.org/10.1080/14616688.2021.2021978.

Tsaur, S.-H., Hsu, F.-S. and Lin, H. (2019) Workplace fun and work engagement in tourism and hospitality: The role of psychological capital. *International Journal of Hospitality Management* 81, 131–140.

Tucker, H. and Shelton, E.J. (2018) Tourism, mood and affect: Narratives of loss and hope. *Annals of Tourism Research* 70, 66–75.

Vada, S., Prentice, C., Scott, N. and Hsiao, A. (2020) Positive psychology and tourist well-being: A systematic literature review. *Tourism Management Perspectives* 33, 100631. https://doi.org/10.1016/j.tmp.2019.100631.

Part 1
Theoretical and Philosophical Foundations

1 Moving from Positive Psychology to Positive Tourism: A Conceptual Approach

Metin Kozak

Introduction

Although there has been little empirical research on positive tourism, it has been central to recent tourism studies. Filep *et al.* (2017) noted that the idea of positive tourism comes from the philosophy of humanism, which dates to the 14th century and has even earlier roots in Ancient Greek philosophy from Aristotle's works on virtues and the good life. This was used as the foundation to establish the roots of psychology in the 20th century, leading first to humanistic psychology and later to positive psychology. Positive psychology is simply considered as 'the study of what makes life worth living' (Filep *et al.*, 2017: 5), which includes a broad focus on health, leisure and recreation, including sports, tourism, and well-being.

The idea also complies with Maslow's (1954) hierarchy of needs, which states the specific steps individuals are expected to take in order to enrich their overall quality of life and well-being, thus seeking to achieve their full potential. According to the hierarchy of needs – which begins at the bottom of a pyramid and stretches to the top – in order to make life worth living people need to partially or fully satisfy their varying needs, including physical (e.g. safety, security, hunger) and psychological (e.g. love, respect, self-esteem, aesthetic) needs. The accumulation of these outcomes refers to *self-actualisation* (at the top of the model pyramid). The degree of influence of such needs varies based on a person's expectations, which result from their social, economic and psychological background.

Despite the existence of a great deal of work on positive psychology, there has been limited emphasis on 'positive tourism', specifically in the field of tourism studies (e.g. Filep *et al.*, 2017; Nawijn, 2016). Furthermore, most of the existing positive psychology/positive tourism works have

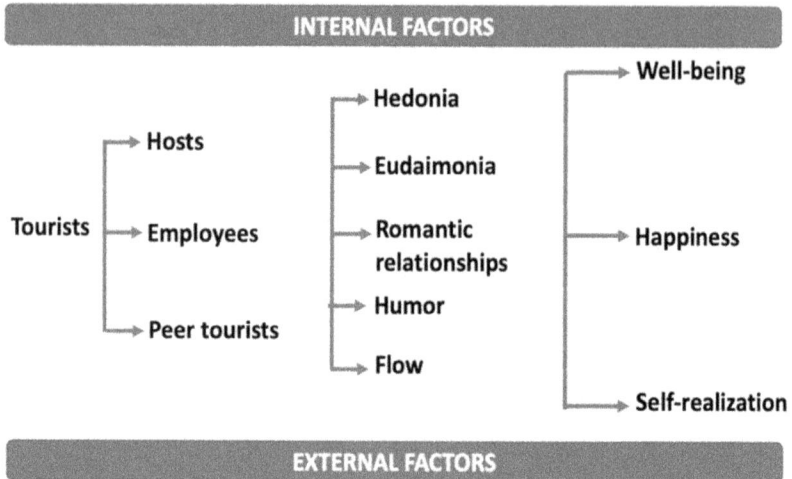

Figure 1.1 Elements of positive tourism
Source: The author's own elaboration.

focused on tourist behaviour and have had limited focus on employees and local communities. As a result, this chapter aims first to delineate the meaning of positive tourism and then to emphasise its importance as a holistic approach. Finally, as the success of positive tourism is central to constructive outcomes of interactions among various parties in the tourism system, we may argue that a diagram to showcase how the elements of positive tourism can be positioned at both the micro and macro levels (see Figure 1.1).

Conceptual Considerations

Tourism, in its modern setting, has been considered a multidisciplinary field of study because its main focus lies at the centre of human beings (Kozak & Kozak, 2016; Tribe & Xiao, 2011; Xiao & Smith, 2006). Since its foundation as an academic field of study, tourism has been heavily influenced by several major fields, such as sociology, anthropology, geography and psychology (Kozak & Kozak, 2016). Of these, psychology looks at the aspects that tourists consider when making travel decisions and how a holiday may help them improve their state of mind, while sociology sits at the centre of how they may interact with other visitors, employees, or residents while on holiday. All these experiences directly and indirectly lead to the well-being of humans – as visitors, employees and residents. This is the mission of positive psychology.

The new millennium has been noted to have consumerism at its centre. Over the last two decades, since the introduction of the concept of the experience economy by Pine and Gilmore (1998) and its subsequent continuation by Vargo and Lusch (2004), there has been a great deal of research regarding the application of service experience in a tourism and hospitality context. These studies have usually paid attention to investigating motivations, emotions, and experiences and how these may impact the well-being and/or quality of life of visitors, employees and hosts. Other subjects such as hedonia, eudaimonia, romantic relationships, humour, and flow have attracted little attention, despite their importance. Figure 1.1 exhibits how the elements of positive tourism can be positioned, together with the influence of internal and external factors. Each of these elements' association with tourism is given in detail below.

Tourism and hedonia

In *Authentic Happiness*, Martin Seligman (2002), one of the eminent voices of positive psychology, explored the subject of life satisfaction, which is synonymous with the concept of 'psychological happiness' used by philosophers. The contents of most current studies on positive psychological well-being have been based on the examination of two broad philosophical traditions regarding the meaning of the quality of life (Aristotle, 2014; Diener *et al.*, 1999; Feldman, 2004; Ryan & Deci, 2001; Ryff & Keyes, 1995). Among these traditions, hedonic well-being seeks 'happiness', while the eudemonic tradition expresses personal development and a meaningful life (Voigt, 2017) and utilitarianism refers to forming group happiness or the greatest good for the greatest number of others, as well as one's own good (Crisp, 1997).

The Greek philosopher Aristippus (435–355 BC) – one of the most important defenders of hedonism – argued that feelings related to pleasure are the best good in life and that all living things, by their nature, seek pleasure and avoid pain (Feldman, 2004). This has been expressed through the concept of the *summum bonum* or pleasure as the supreme good in life that was first introduced by Cicero (106–43 BC). According to the hedonic approach, individuals experience positive emotions frequently and negative emotions less frequently, and receive high satisfaction from their life (Diener, 1984). Equivalent to pleasure and happiness, hedonic well-being (Kahneman *et al.*, 1999) entails seeking pleasure (positive and pleasant emotions) and comfort (relaxation and ease) (Huta & Ryan, 2010). For example, the tourism literature suggests that seeking pleasure and relaxation is among the highly desired push motivations involved in making decisions to travel (Güzel *et al.*, 2020; Kozak, 2002; Pestana *et al.*, 2020).

According to Huta and Ryan (2010), hedonia is generally related to feeling good during activities. Eudaimonia is concerned with the sense that one's activities are meaningful and valuable. In other words, even if a person leaves a challenging activity with negative emotions, the resulting increased skill level will cause the person to experience meaningful emotions, such as self-actualisation, promotion to a higher level of functionality, feeling complete, and experiencing competence in important areas of life. In short, it is not possible to classify emerging behaviours as hedonic or eudemonic, based on the fact that each individual has different motivations surrounding an experience or activity. As the resulting effects vary from individual to individual, the reflections of the effects also differ.

Tourism and eudaimonia

The philosophical roots of positive psychology (Seligman & Csikszentmihalyi, 2000) express the examination of what makes life worth living via a simple definition: at the centre of this is the concept of eudaimonia, which Aristotle described as happiness and good life (Voigt, 2017). The foundation of eudaimonia, which is based on a humanistic idea that strives towards perfection, is a higher developmental state that emerges as a result of the individual's self-development and self-actualisation (Cloninger, 2004; Ryff & Singer, 1996). In order to better understand eudemonic well-being from a conceptual point of view, it is useful to examine Aristotle's *Nichomachean Ethics*. In this work, Aristotle (2014) stated that eudaimonia is at the top of everything that human beings can reach or achieve. The goal for human beings includes the satisfaction of all criteria for the highest good. A person experiences happiness as a result of self-realisation; thus, happiness is a goal that all people aim to pursue by engaging in various activities such as leisure, sports, socialisation and gastronomy.

The basic dimensions of eudemonic well-being appear in Ryff's (1989: 1071) conceptualisation of 'psychological well-being', which includes six dimensions: self-acceptance, positive relations with others, autonomy, environmental mastery, purpose in life and personal growth. A recent, alternative approach to eudemonic well-being was posited by Huta and Waterman (2013: 1435) and introduced four eudemonic dimensions: growth (self-actualisation/self-improvement/maturity), authenticity (identity/personal expression/autonomy/constitutive goals/integrity), meaning (determining the purpose in life/long-term perspective/contributing) and excellence (virtue/using the best of oneself/reaching a high standard).

Both types of well-being can be usefully combined as an antecedent, outcome, and mediator variable and then interpreted together (Keyes *et al*., 2002). According to Huta and Ryan (2010), activities with hedonic motivations are more associated with instant well-being; activities with eudemonic motivations are associated with longer-term well-being. However, when the source of motivation is both, the hedonic and eudemonic motivated states are interrelated; both are necessary for optimal psychological well-being.

Efforts to use positive psychology to examine the links between tourism and well-being have increased as a result of the easy scientific evaluation of previously obscure structures from positive psychology (Filep, 2014; Pearce, 2009b). The relationship between tourism and well-being has gradually become one of the main topics in current tourism publications, and research conducted on the well-being of different stakeholders examines the assumed benefits and costs of tourism.

Tourism studies have begun to focus on well-being inspired by psychology and philosophy, for example, 'quality of life' and 'life satisfaction/ satisfaction' (e.g. de Bloom *et al.*, 2010; Dolnicar *et al.*, 2012; Hoopes & Lounsbury, 1989; Liu, 2015), 'happiness' (e.g. Bimonte & Faralla, 2013; Filep, 2014; McCabe & Johnson, 2013; Nawijn, 2010), 'wellness' (e.g. Kelly, 2010; Luo *et al.*, 2017; Smith & Kelly, 2006; Smith & Puczkó, 2009; Voigt *et al.*, 2011) and *summum bonum* (e.g. Fennell, 2018). However, Filep (2014) stated that almost all existing empirical studies in tourism have focused only on hedonic well-being. Emotions related to pleasure (hedonistic feelings) are a strong motivator to participate in a tourism activity (Crompton, 1979; Fennell, 2009; Fodness, 1994; Singh, 2015). However, more inclusive well-being studies that include eudaimonia have recently been increasing (e.g. Ahn *et al.*, 2019; Cai *et al.*, 2020; Lee & Jeong, 2019; Su *et al.*, 2020).

Today, consumers' emotions have become an important factor in evaluating their experiences and behavioural responses due to their increasing experiences and interactions with other consumers (Hosany *et al.*, 2015; Pestana *et al.*, 2020; Şanlıöz-Özgen & Kozak, 2021; Volo, 2021). Hedonic feelings are an important factor in determining the satisfaction level of tourists (Kim *et al.*, 2010). However, measuring the happiness of tourists according to the pleasure they receive and ignoring meaningful experiences that add value to them is insufficient (Filep, 2014). Therefore, it is too narrow to focus only on hedonic effects and to ignore the effects that can be caused by meaningful experiences (Knobloch *et al.*, 2016).

Tourism and romantic relationships

As indicated by Maslow (1954), the need to feel a sense of belonging and connection to others, as well as to feel valued by those to whom one feels connected, is an innate basic human need and is necessary for well-being (Diener & Fujita, 1995; Ryan & Deci, 2000). The support of the individual in social relations is an important factor affecting the level of subjective well-being because trust in close social relationships increases happiness. When happy and unhappy people have been compared, happy people have been observed to be more social, have stronger and more satisfying relationships, be more extroverted, have more success in interpersonal relationships, experience more positive than negative emotions and be able to more easily overcome negative events in a short time (Biswas-Diener *et al.*, 2004; Diener & Seligman, 2002; Myers & Diener, 1995). The involvement in several forms of leisure, recreation and tourism activities can be regarded as the mediating factor in leading to happiness and maintaining its continuity in one's life cycle.

In order to meet an individual's need for attention, spending free time with their partners is important in terms of establishing and strengthening positive social relations (Ryan *et al.*, 2010). Leisure time (outside of work hours and perhaps only available on weekends) is a time period traditionally associated with independent and close relationships (Rybczynski, 1991) as it is limited in terms of providing opportunities to spend an extended period of time with significant others. Longer and uninterrupted free time can be an important opportunity for people to be with their loved ones and to consolidate social ties as a result of positive interactions (de Bloom *et al.*, 2017).

The empirical evidence suggests that holidays with family have a positive effect on family functioning, as tourism experiences allow the sharing of memories, improve communication within the family, and ultimately reinforce feelings of togetherness and harmony (Lehto *et al.*, 2012; Shaw & Dawson, 2001). Since leisure activities conducted together cause a positive emotional increase in relationship dynamics, this indicates an important link between leisure time and well-being (de Bloom *et al.*, 2017; Ryan & Deci, 2000). Many studies in the field of leisure have shown that couples and families enjoy increased satisfaction in their relationships or marital life when they spend some free time together (e.g. Durko & Petrick, 2013; Johnson *et al.*, 2006; Kelly & Kelly, 1994).

In addition, as tourism allows a temporary change of place and opportunities to meet new faces, romantic relationships may also result from interactions with hosts, employees or fellow tourists while on holiday. The choice of destinations would also be relevant in this context as different locations relate to the search for sex, love, or companionship (Bauer & McKercher, 2003) – for instance, seeking solitude and privacy or searching for partners by choosing hedonic tourism destinations that offer sexual freedom such as Las Vegas, Bangkok or Daytona Beach. This may last for either the short or the long term, depending on the expectations or influence of situational factors such as distance, culture, and so on. Despite negative reactions and ongoing debates on its ethical consequences, sex tourism may also be a form of positive psychology in a one-way or asynchronous relationship. Finally, the study by de Bloom *et al.* (2017) has found empirical evidence supporting the suggestion that tourism plays a mediating role in increasing the happiness of couples due to their more intense romantic relationships during the holiday.

Tourism and humour

Generally speaking, humour is a multidisciplinary subject studied in various fields, such as psychology, sociology and anthropology (Raskin, 1987). In the context of psychology, Ruch (1995) stated that, although emotions cannot be observed directly, they are hypothetical inferences made through various indicators. These inferences are based on the individual's behaviour, physiological changes or subjective experiences. The emotion elicited by the humour stimuli is the real positive emotion expressed. Positive psychology emphasises the importance of understanding not only negative moods and stress factors but also how to be happy (Seligman et al., 2005).

Thus, all processes of human experience are explored and understood through positive psychology, which means finding ways to make life enjoyable, worth living, and virtuous (Peterson & Seligman, 2004; Seligman & Csikszentmihalyi, 2000). This may depend upon the person's needs and expectations in making the decision to travel. For example, an allocentric type of person may enrich their enjoyment by getting involved in adventure-based leisure activities such as hiking, rafting, sailing or hunting, while visiting holy places such as Mecca or the Vatican may give travellers the feeling that life is worth living. Humour – which includes radiating joy and the ability to make fun of one's own embarrassing moments – can have positive effects on life satisfaction (Beermann & Ruch, 2009). Laughter, humour and response to humour have the potential to improve mood and the resulting positive emotions are beneficial for the health of human beings (Fredrickson, 1998; Solomon, 1996).

There is also a close relationship between tourism and humour (Cohen, 2011). Tourism businesses use humour to make holidays with family and close friends, which are considered a pleasant activity, restful and entertaining – and thus more enjoyable – in order to increase people's well-being (Pabel, 2017). Pearce (2009a) suggested that humour has multiple uses in tourism environments (such as humour for, about, and by tourists) and that these roles and outcomes are realised and achieved through various mechanisms. The role of tourism humour is especially important in ensuring tourists' comfort, helping to increase concentration (expanding people's attention and capturing their level of interest), and establishing social connections (bringing people together) by connecting tourists and hosts.

However, it is important to reveal tourists' perspective about such humour, as it contributes to the quality of the tourism experience only if it is used consciously and appropriately (Pabel, 2017). Considering that businesses operating in the tourism industry are trying to attract attention for their own brands, services, and products, attention can be drawn by appealing to the audience's emotions or to their sense of humour (Hoyer & MacInnis, 2008). The degree of familiarity between tourists and hosts also directly and indirectly contributes to the quality of humour.

Pearce and Pabel (2014) stated that humour can be used effectively to promote destinations and make attractions and activities exciting. In some cases, humour can help tourists cope with awkward and embarrassing situations they find themselves in. There are, however, some difficulties when using humour in tourism, which is a social platform that serves people from different nationalities and cultural groups. One of these difficulties lies in making the humour seem natural; another lies in using it in a moderate and respectful manner, without disturbing the tourist or the target of the humour (Pabel & Pearce, 2016). When developing a tourism experience through humour, the ability to anticipate the audience's reaction and obtain feedback is critical in creating enjoyable and entertaining presentations (Pabel & Pearce, 2018).

Tourism and flow

Flow theory was developed in studies on the creativity of artists, athletes and others who use creativity in their work (Csikszentmihalyi, 1975, 1990, 1997). The concept of flow was defined by Csikszentmihalyi (1975) as a psychological state – a high level of performance and a very positive experience resulting from people's perceptions of difficulties and skills in certain situations. According to Filep (2008), flow is the outcome of skill–challenge balance and travel experiences lead to tourist satisfaction through flow.

Cohen (1979) characterised recreation as an entertainment activity because recreational activities provide tourists with a general sense of well-being and a sense of pleasure at the same time. Panchal (2014) stated that it is possible to associate this situation with the concept of flow because in settings where the pleasure from experiences is dominant, the flow expressed by a high-achievement or positive experience becomes optimal, as expressed in positive psychology.

During the flow experience, it is necessary for the person to give feedback about the path taken to reach the goal; action and awareness should be combined in such a way that the feedback will be a part of the strategic plan for the next step toward the goal and the person should overcome any fear of failure (Nakamura & Csikszentmihalyi, 2002). For example, while often during an experience time seems to have slowed down, since control is entirely in the hands of the individual and in general they are focusing on the goal, they may later be surprised to find out how fast time actually passed (Csikszentmihalyi, 1990). People can also experience holidays that end more quickly than expected, without being aware of how time seemed to fly. Experiencing flow during such a holiday can be much more rewarding than having a holiday full of obstacles (Csikszentmihalyi, 1997; Csikszentmihalyi & Coffey, 2017).

Conclusion

Over the years, tourism has become an important part of human beings' modern life. External factors (e.g. migration and urbanisation) and internal factors (e.g. stress, loneliness) lead people to participate in tourism activities.

For this reason, as in other fields, there is a close relationship between tourism and psychology. If we consider the mutual relationship between these two fields, while a person participating in tourism feels good psychologically, psychology also points to tourism in the context of feeling good and reaching happiness. Whether it is for health, cultural, religious or recreational purposes, people participate in tourism because they want to achieve happiness from feeling good. Positive psychology deals with the relationship with happiness in the context of tourism. There are two major approaches to reaching happiness through positive psychology: hedonic (subjective well-being) and eudaimonic (psychological well-being).

This chapter has looked at the association between positive psychology and tourism in the specific context of hedonia, eudaimonia, romantic relationships, humour and flow. It has further elaborated on how the outputs of this association may lead to well-being, happiness or self-realisation during a holiday and/or its following periods. As explicitly discussed within this chapter, going on a holiday brings with it many positive psychological and physiological aspects of well-being, such as better health, improved social and romantic relationships, positive impacts, higher job satisfaction and feelings of renewal.

The rewarding nature of experiencing hedonia, eudaimonia, romantic relationships, humour, and flow while on holiday can motivate people to stay in a state of flow after the holiday is over, and continuation of the experiences obtained during the holiday can result in similar well-being, happiness or self-realisation. For example, when an individual takes a holiday filled with any of the above-mentioned experiences, they may return to their life in a renewed state, achieving a higher level of focus and positive results that extend to both their private and work life.

References

Ahn, J., Back, K.-J. and Boger, C. (2019) Effects of integrated resort experience on customers' hedonic and eudaimonic well-being. *Journal of Hospitality and Tourism Research* 43 (8), 1–31.
Aristotle (2014) *Nicomachean Ethics*. Translated by C.D.C. Reeve. Indianapolis and Cambridge: Hackett Publishing Company.
Bauer, T. and McKercher, B. (2003) *Sex and Tourism: Journeys of Romance, Love, and Lust*. Binghamton: The Haworth Hospitality Press.
Beermann, U. and Ruch, W. (2009) How virtuous is humor? Evidence from everyday behavior. *Humor – International Journal of Humor Research* 22 (4), 395–417.
Bimonte, S. and Faralla, V. (2013) Happiness and outdoor vacations appreciative versus consumptive tourists. *Journal of Travel Research* 54 (2), 179–192.
Biswas-Diener, R., Diener, E. and Tamir, M. (2004) The psychology of subjective wellbeing. *Daedalus* 133 (2), 18–25.
Cai, Y., Ma, J. and Lee, Y.-S. (2020) How do Chinese travelers experience the Arctic? Insights from a hedonic and eudaimonic perspective. *Scandinavian Journal of Hospitality and Tourism* 20 (2), 144–165.
Cloninger, C.R. (2004) *Feeling Good: The Science of Well-Being*. New York: Oxford University Press.

Cohen, E. (1979) A phenomenology of tourist experiences. *Sociology* 13 (2), 179–201.
Cohen, E. (2011) The people of tourism cartoons. *Anatolia: An International Journal of Tourism and Hospitality Research* 22 (3), 326–349.
Crisp, R. (1997) *Mill on Utilitarianism*. London: Routledge.
Crompton, J.L. (1979) Motivations for pleasure vacation. *Annals of Tourism Research* 6 (4), 408–424.
Csikszentmihalyi, M. (1975) *Beyond Boredom and Anxiety*. San Francisco: Jossey-Bass.
Csikszentmihalyi, M. (1990) *Flow: The Psychology of Optimal Experience*. New York: Harper & Row.
Csikszentmihalyi, M. (1997) *Finding Flow: The Psychology of Engagement with Everyday Life*. New York: Basic Books.
Csikszentmihalyi, M. and Coffey, J.K. (2017) Why do we travel? A positive psychological model for travel motivation. In S. Filep, J. Laing and M. Csikszentmihalyi (eds) *Positive Tourism* (pp. 122–132). Abingdon: Routledge.
De Bloom, J., Geurts, S.A.E., Taris, T.W., Sonnentag, S., de Weerth, C. and Kompier, M.A.J. (2010) Effects of vacation from work on health and well-being: Lots of fun, quickly gone. *Work & Stress* 24 (2), 196–216.
De Bloom, J., Geurts, S. and Lohmann, M. (2017) Tourism and love: How do tourist experiences affect romantic relationships? In S. Filep, J. Laing and M. Csikszentmihalyi (eds) *Positive Tourism* (pp. 35–53). Abingdon: Routledge.
Diener, E. (1984) Subjective well-being. *Psychological Bulletin* 95, 542–575.
Diener, E. and Fujita, F. (1995) Resources, personal strivings, and subjective well-being: A nomothetic and idiographic approach. *Journal of Personality and Social Psychology* 68 (5), 926–935.
Diener, E. and Seligman, M.E.P. (2002) Very happy people. *Psychological Science* 13 (1), 81–84.
Diener, E., Suh, E.M., Lucas, R.E. and Smith, H.L. (1999) Subjective well-being: Three decades of progress. *Psychological Bulletin* 125 (2), 276–302.
Dolnicar, S., Yanamandram, V. and Cliff, K. (2012) The contribution of vacations to quality of life. *Annals of Tourism Research* 39 (1), 59–83.
Durko, A.M. and Petrick, J.F. (2013) Family and relationship benefits of travel experiences. *Journal of Travel Research* 52 (6), 720–730.
Feldman, F. (2004) *Pleasure and the Good Life: Concerning the Nature, Varieties, and Plausibility of Hedonism*. New York: Oxford University Press.
Fennell, D.A. (2009) The nature of pleasure in pleasure travel. *Tourism Recreation Research* 34 (2), 123–134.
Fennell, D.A. (2018) On tourism, pleasure and the *summum bonum*. *Journal of Ecotourism* 17 (4), 383–400.
Filep, S. (2008) Applying the dimensions of flow to explore visitor engagement and satisfaction. *Visitor Studies* 11 (1), 90–108.
Filep, S. (2014) Moving beyond subjective well-being: A tourism critique. *Journal of Hospitality & Tourism Research* 38 (2), 266–274.
Filep, S., Laing, J. and Csikszentmihalyi, M. (eds) (2017) *Positive Tourism*. Abingdon: Routledge.
Fodness, D. (1994) Measuring tourist motivation. *Annals of Tourism Research* 21 (3), 555–581.
Fredrickson, B.L. (1998) What good are positive emotions? *Review of General Psychology* 2 (3), 300–319.
Güzel, Ö., Sahin, I. and Ryan, C. (2020) Push-motivation-based emotional arousal: A research study in a coastal destination. *Journal of Destination Marketing & Management* 16, 100428.
Hoopes, L.L. and Lounsbury, J.W. (1989) An investigation of life satisfaction following a vacation: A domain-specific approach. *Journal of Community Psychology* 17 (2), 129–140.

Hosany, S., Prayag, G., Deesilatham, S., Causevic, S. and Odeh, K. (2015) Measuring tourists' emotional experiences: Further validation of the destination emotion scale. *Journal of Travel Research* 54 (4), 482–495.

Hoyer, W. and MacInnis, D. (2008) *Consumer Behavior*. Mason: South-Western College.

Huta, V. and Ryan, R.M. (2010) Pursuing pleasure or virtue: The differential and overlapping well-being benefits of hedonic and eudaimonic motives. *Journal of Happiness Studies* 11 (6), 735–762.

Huta, V. and Waterman, A.S. (2013) Eudaimonia and its distinction from hedonia: Developing a classification and terminology for understanding conceptual and operational definitions. *Journal of Happiness Studies* 15 (6), 1425–1456.

Johnson, H.A., Zabriskie, R.B. and Hill, B. (2006) The contribution of couple leisure involvement, leisure time, and leisure satisfaction to marital satisfaction. *Marriage & Family Review* 40 (1), 69–91.

Kahneman, D., Diener, E. and Schwarz, N. (eds) (1999) *Well-Being: Foundations of Hedonic Psychology*. New York: Russell Sage Foundation.

Kelly, C. (2010) Analysing wellness tourism provision: A retreat operators' study. *Journal of Hospitality and Tourism Management* 17 (1), 108–116.

Kelly, J.R. and Kelly, J.R. (1994) Multiple dimensions of meaning in the domains of work, family, and leisure. *Journal of Leisure Research* 26 (3), 250–274.

Keyes, C.L.M., Shmotkin, D. and Ryff, C.D. (2002) Optimizing well-being: The empirical encounter of two traditions. *Journal of Personality and Social Psychology* 82 (6), 1007–1022.

Kim, J.-H., Ritchie, J.R.B. and Tung, V.W.S. (2010) The effect of memorable experience on behavioral intentions in tourism: A structural equation modeling approach. *Tourism Analysis* 15 (6), 637–648.

Knobloch, U., Robertson, K. and Aitken, R. (2016) Experience, emotion, and eudaimonia: A consideration of tourist experiences and well-being. *Journal of Travel Research* 56 (5), 651–662.

Kozak, M. (2002) Comparative analysis of tourist motivations by nationality and destinations. *Tourism Management* 23 (3), 221–232.

Kozak, M. and Kozak, N. (2016) Institutionalisation of tourism research and education: From the early 1900s to 2000s. *Journal of Tourism History* 8 (3), 275–299.

Lee, W. and Jeong, C. (2019) Beyond the correlation between tourist eudaimonic and hedonic experiences: Necessary condition analysis. *Current Issues in Tourism* 23 (17), 2182–2194.

Lehto, X.Y., Lin, Y.-C., Chen, Y. and Choi, S. (2012) Family vacation activities and family cohesion. *Journal of Travel & Tourism Marketing* 29 (8), 835–850.

Liu, Y.-D. (2015) Event and quality of life: A case study of Liverpool as the 2008 European Capital of Culture. *Applied Research in Quality of Life* 11 (3), 707–721.

Luo, Y., Lanlung (Luke), C., Kim, E., Tang, L.R. and Song, S.M. (2017) Towards quality of life: The effects of the wellness tourism experience. *Journal of Travel & Tourism Marketing* 35 (4), 410–424.

Maslow, A.H. (1954) *Motivation and Personality*. New York: Harpers and Row.

McCabe, S. and Johnson, S. (2013) The happiness factor in tourism: Subjective well-being and social tourism. *Annals of Tourism Research* 41, 42–65.

Myers, D.G. and Diener, E. (1995) Who is happy? *Psychological Science* 6 (1), 10–19.

Nakamura, J. and Csikszentmihalyi, M. (2002) The concept of flow. In C.R. Snyder and S.J. Lopez (eds) *Oxford Handbook of Positive Psychology* (pp. 89–105). New York: Oxford University Press.

Nawijn, J. (2010) The holiday happiness curve: A preliminary investigation into mood during a holiday abroad. *International Journal of Tourism Research* 12 (3), 281–290.

Nawijn, J. (2016) Positive psychology in tourism: A critique. *Annals of Tourism Research* 56, 151–153.

Pabel, A. (2017) The role of humour in contributing to tourism experiences. In S. Filep, J. Laing and M. Csikszentmihalyi (eds) *Positive Tourism* (pp. 86–104). Abingdon: Routledge.

Pabel, A. and Pearce, P.L. (2016) Humor. In J. Jafari and H. Xiao (eds) *Encyclopedia of Tourism* (pp. 445–446). Cham: Springer.

Pabel, A. and Pearce, P.L. (2018) Selecting humour in tourism settings – A guide for tourism operators. *Tourism Management Perspectives* 25, 64–70.

Panchal, J. (2014) Tourism, wellness and feeling good: Reviewing and studying Asian spa experiences. In S. Filep and P. Pearce (eds) *Tourist Experience and Fulfilment: Insights from Positive Psychology* (pp. 72–87). Abingdon: Routledge.

Pearce, P.L. (2009a) Now that is funny: Humour in tourism settings. *Annals of Tourism Research* 36 (4), 627–644.

Pearce, P.L. (2009b) The relationship between positive psychology and tourist behavior studies. *Tourism Analysis* 14 (1), 37–48.

Pearce, P. and Pabel, A. (2014) Humour, tourism and positive psychology. In S. Filep and P. Pearce (eds) *Tourist Experience and Fulfilment: Insights from Positive Psychology* (pp. 17–36). Abingdon: Routledge.

Pestana, M.H., Parreira, A. and Moutinho, L. (2020) Motivations, emotions and satisfaction: The keys to a tourism destination choice. *Journal of Destination Marketing & Management* 16, 100332.

Peterson, C. and Seligman, M.E.P. (2004) *Character Strengths and Virtues: A Handbook and Classification*. New York: Oxford University Press.

Pine, B.J. and Gilmore, J.H. (1998) Welcome to the experience economy. *Harvard Business Review*, July–August, 98–105.

Raskin, V. (1987) Linguistic heuristics of humor: A script-based semantic approach. *International Journal of the Sociology of Language* 65, 1–25.

Ruch, W. (1995) Will the real relationship between facial expression and affective experience please stand up: The case of exhilaration. *Cognition & Emotion* 9 (1), 33–58.

Ryan, R.M. and Deci, E.L. (2000) Self-determination theory and the facilitation of intrinsic motivation, social development, and well-being. *American Psychologist* 55, 68–78.

Ryan, R.M. and Deci, E.L. (2001) On happiness and human potentials: A review of research on hedonic and eudaimonic well-being. *Annual Review of Psychology* 52 (1), 141–166.

Ryan, R.M., Bernstein, J.H. and Brown, K.W. (2010) Weekends, work, and well-being: Psychological need satisfactions and day of the week effects on mood, vitality, and physical symptoms. *Journal of Social and Clinical Psychology* 29 (1), 95–122.

Rybczynski, W. (1991) *Waiting for the Weekend*. New York: Viking Penguin.

Ryff, C.D. (1989) Happiness is everything, or is it? Explorations on the meaning of psychological well-being. *Journal of Personality and Social Psychology* 57 (6), 1069–1081.

Ryff, C.D. and Keyes, C.L.M. (1995) The structure of psychological well-being revisited. *Journal of Personality and Social Psychology* 69 (4), 719–727.

Ryff, C.D. and Singer, B. (1996) Psychological well-being: Meaning, measurement, and implications for psychotherapy research. *Psychotherapy and Psychosomatics* 65 (1), 14–23.

Şanlıöz-Özgen, H.K. and Kozak, M. (2021) The interrelationship between tourist satisfaction and experiences: How does one contribute to the other? In R.A. Sharpley (ed.) *The Routledge Handbook of Tourist Experience* (pp. 64–76). New York: Routledge.

Seligman, M.E.P. (2002) *Authentic Happiness: Using the New Positive Psychology to Realize Your Potential for Lasting Fulfillment*. New York: The Free Press.

Seligman, M.E.P. and Csikszentmihalyi, M. (2000) Positive psychology: An introduction. *American Psychologist* 55 (1), 5–14.

Seligman, M.E.P., Steen, T.A., Park, N. and Peterson, C. (2005) Positive psychology progress: Empirical validation of interventions. *American Psychologist* 60 (5), 410–421.

Shaw, S.M. and Dawson, D. (2001) Purposive leisure: Examining parental discourses on family activities. *Leisure Sciences* 23 (4), 217–231.

Singh, T.V. (ed.) (2015) *Challenges in Tourism Research*. Bristol: Channel View Publications.

Smith, G.T., Snyder, D.K., Trull, T.J. and Monsma, B.R. (1988) Predicting relationship satisfaction from couples' use of leisure time. *The American Journal of Family Therapy* 16 (1), 3–13.

Smith, M. and Kelly, C. (2006) Journeys of the self: The rise of the wellness tourism sector. *Journal of Tourism and Recreation Research* 31 (1), 15–25.

Smith, M.K. and Puczkó, L. (2009) *Health and Wellness Tourism*. Oxford: Butterworth Heinemann-Elsevier.

Solomon, J.C. (1996) Humor and aging well: A laughing matter or a matter of laughing? *American Behavioral Scientist* 39 (3), 249–271.

Su, L., Tang, B. and Nawijn, J. (2020) Eudaimonic and hedonic well-being pattern changes: Intensity and activity. *Annals of Tourism Research* 84, 1–14.

Tribe, J. and Xiao, H. (2011) Developments in tourism social science. *Annals of Tourism Research* 38 (1), 7–26.

Vargo, S. and Lusch, R. (2004) Evolving to a new dominant logic for marketing. *Journal of Marketing* 68 (1), 1–17.

Voigt, C. (2017) Employing hedonia and eudaimonia to explore differences between three groups of wellness tourists on the experiential, the motivational and the global level. In S. Filep, J. Laing and M. Csikszentmihalyi (eds) *Positive Tourism* (pp. 105–121). Abingdon: Routledge.

Voigt, C., Brown, G. and Howat, G. (2011) Wellness tourists: In search of transformation. *Tourism Review* 66 (1/2), 16–30.

Volo, S. (2021) The experience of emotion: Directions for tourism design. *Annals of Tourism Research* 86, 103097. https://doi.org/10.1016/j.annals.2020.103097.

Xiao, H. and Smith, S.L.J. (2006) The making of tourism research: Insights from a social sciences journal. *Annals of Tourism Research* 33 (2), 490–507.

2 The Search for Meaning in Life: The Role of Tourism

Carla Fraga and Vera Lúcia Bogéa Borges

Introduction

Post-Fordist tourism is gathering strength, focusing on contemplation and the search for meaning in life. The tourist market is therefore increasingly organised around customised experiences according to the demands of modern tourists, who are motivated by the need for sustainability according to the sustainable development goals (SDGs) proposed by the United Nations (UN, 2015). As a result, efforts to attract visitors to destinations are aimed at the formation of specific niches; slow tourism and yoga tourism are but two of many.

However, since the start of 2020, the COVID-19 pandemic has resulted in social distancing, travel restrictions, and lockdowns in various tourist destinations, as well as other measures to contain transmission, negatively impacting tourism worldwide. For instance, considering global tourism in the context of economic transformation, Prideaux *et al.* (2020: 667) highlight the relevance of lessons learned from COVID-19: for example, the tourist industry should 'seek to understand how it should respond to the emerging transformation of the global economy to carbon neutrality'. From the point of view of tourism demand and supply – businesses, destinations and policymakers – Sigala (2021: 319) takes us through the stages of the impacts and implications of COVID-19: (1) response, (2) recovery and (3) reset, and the answers to the inflexion question, 'What is more and what is next?'

With the start of vaccination campaigns, hopes rose of a return to pre-pandemic tourism patterns, but predicting the post-pandemic scenario will require an understanding of human behaviour in reaction to anxieties, since social pains (i.e. marks left from past experiences and present reality) have been and are still being generated by this challenging moment in human history. In this respect, interfaces with other areas of knowledge, such as neurosciences and applied psychology, can provide a path to better understanding the current tourism phenomenon.

This chapter is divided into three main parts (excluding this introduction and the final considerations). In the first part, we briefly describe the history of neurosciences and their relevance to the study of the tourism phenomenon. In the second, we focus on the topic of social pains, starting with a specific examination of social ostracism based on knowledge from neurosciences and applied psychology. In the third part, we reflect more deeply on human motivations (instinct theory, drive reduction theory, optimal arousal theory, hierarchy of needs theory, and so on) as powerful elements for repositioning tourist destinations in the period after the pandemic. As an exercise in future perspectives, the study contributes to the field by applying principles from multiple disciplines to understanding the tourism phenomenon and its epistemological developments in reaction to perhaps the greatest challenge of the 21st century so far, the COVID-19 pandemic. Thus, the purpose of the chapter relates to the meaning of life and tourism, because tourist experiences are transformative from a cognitive and behavioural point of view (see Scott, 2020).

Neurosciences and the Tourism Phenomenon

> Advancements in the biology of the mind offer the possibility of a new humanism, one that mixes sciences, those that are preoccupied with the natural world, with humanities, i.e. those that concern themselves with the significance of the human experience. (Kandel, 2020: xiv; our translation)

Tourism is a complex phenomenon that involves the sum of a series of moments that constitute the tourism experience, this being exactly where the relationship between neurosciences and tourism begins. It is through this relationship, which are essentially interdisciplinary, that it becomes possible to comprehend the tourism phenomenon (and its various stakeholders, including tourists, local communities, and others) through theories and methods relevant to the in-depth analysis of behaviour and cognition.

In different time periods, each of the various fields of knowledge has acquired its own characteristics by carrying out specific epistemologies that manifest in a particular form of seeing and reading the world, via the use of a specific set of language and techniques. In this sense, specialties have been formed by ways of thinking, ways of giving shape to each object of study, thus making them intelligible (Gurgel, 2017). For a long time, in relation to the human body, the brain was not very well understood, and it seemed to be surrounded by questions regarding the smallest of structures (which cannot be observed without the employment of proper equipment) and their functioning. Therefore, the rise of image techniques suited for the mapping of this crucial part of the human body has allowed the unravelling of many mysteries that encircled it and, from there on, allowed it to be properly studied.

In ancient times, the region of Mesopotamia (currently Iraq, situated within the Tigris–Euphrates river system) was initially occupied by the

Sumerians, and even though there is no certainty regarding the motive for their settling in the region, it is believed that the search for water and food, widely available in the area between the Iran plateau and the Zagros Mountains, was what prompted their abandonment of a nomadic lifestyle (Mella, 2001). By the year 4000 BC, for example, the Sumerians had left records of the euphoria caused in human beings by the seeds of the papoula flower (poppy seeds). In various cultures, the opening of holes in the skull, known as trepanning, was used as a surgical procedure, perhaps with the objective of treating brain disorders such as epilepsy. This conduct could also have been undertaken in the context of rituals and/or spiritual manifestations (Moraes, 2009). In the classical antiquity era, the Greeks seemed to realise that the brain was the centre for mental processes. Plato believed in this, and Aristotle in turn recognised the importance of the heart as a noble organ of the body, although he considered there to be an association between the brain and the heart's temperature (Moraes, 2009; Rooney, 2018).

In modern times, the 19th century was decisive in the advances in various areas that lead to innovation in medicine. In 1889, the history of mankind and neuroscience were forever changed by the proposition of Santiago Ramón y Cajal's neuron doctrine, which highlighted the collateral cells of the white matter, the composition of neurons, and the behaviour of axons and provided a description of the brain's sensory roots, amongst other discoveries registered in various works (Carlos, 2015). According to Azmitia (2015: 100; our translation), 'Cajal recognised the plastic and vital properties of neurons and their contribution to brain functions, such as memory, learning, sleep, and mental disorders. He proposed that diffusible chemical factors influenced the shape of neurons'.

By the rise of the 20th century, various imaging techniques began being used directly or indirectly, arousing the interest of scientists, particularly regarding those related to the brain, which are called neuroimages. According to Rocha *et al.* (2001), several research studies were made that contributed to the understanding of cerebral structures and functioning and, in the case of psychiatric clinics, aided in the treatment of neuropsychiatric disorders. The rise of cognitive neuroscience, which uses even more advanced techniques that allow for 'brain imaging (PET, positron emission tomography, and fMRI, functional magnetic resonance imaging), lead to the study of the cerebral activity in vivo' (Teixeira, 2015: 173; our translation).

The study possibilities have advanced significantly over the last few years. Nowadays, it is possible to comprehend social interactions through hyperscanning, which is a technique that enables the measurement of 'the degree of brain-to-brain synchronizations, termed inter-brain synchrony (IBS)' (Mayseless *et al.*, 2019). Mayseless *et al.* (2019) have advanced an essential topic for teams who want to investigate the interface of tourism and neuroscience: 'real-life creative problem solving in teams'. Although methods such as functional magnetic resonance imaging (fMRI) and electroencephalography (EEG), among others, are utilised for hyperscanning, in their research, Mayseless *et al.* (2019) used functional near-infrared spectroscopy (fNIRS), concluding:

> This study examined processes of innovation in a naturalistic problem-solving paradigm and demonstrated the association between cooperation and positive creative outcomes as well as highlighted the beginning of an interactive brain model of IBS underlying this process. (Mayseless *et al.*, 2019: n.p.)

Therefore, the importance that neuroscience allied with other fields of knowledge such as applied psychology can have for the study of tourism is critical. After a brief historical contextualisation of neuroscience, it is important to illustrate some considerations about its relevance to the study of the tourism phenomenon. In epistemological terms, the paradigmatic vision regarding tourism is the systemic approach, which has its roots in biology – that is, in general systems theory. However, post-paradigmatic approaches are underway (Lohmann & Panosso Netto, 2008) in search of other paradigms in the treatment of this complex phenomenon. Neurotourism seems to be a growing paradigm; for instance, Giudici *et al.* (2017: 340) highlight that '[n]eurotourism is a discipline mixing neuroscience and tourism'. Thus, generally, throughout history, tourism has developed as a cultural response involving the notion of movement and/or dislocation in a world that nowadays is submitted to fast-paced transformation. The alternative possibilities for tourists caused a fragmentation of the market and made it possible for tourism researchers to instigate innovations, questions, creations and reformulations in relation to the production of knowledge (Panosso Netto, 2010).

In the same way, the possibility for dialogue with other areas of knowledge that otherwise seemed impractical makes the interaction between neuroscience and tourism challenging and stimulating. Through this juxtaposition, theories and methods from the neurosciences and from the upcoming field of neurotourism can contribute to the comprehension of aspects that have not been widely discussed before in the scientific literature of tourism. As an example, Tosun *et al.* (2016: 20) analysed consumer preference by deploying methods from neurotourism. The authors used EEG to evaluate choices about hotels; they explain that:

> [c]onsumers reach a decision with the effect of psychological factors including personality, living style, perception, motivation, learning and attitude; demographic factors including age, gender, income level, occupation, educational status and family size; and socio-cultural factors including family, advisory group, social class and culture. There are a series of sub decisions in touristic purchasing behavior including vacation spot, transportation, accommodation, activities, budget and reservation. An issue that needs attention first is to find answers to the questions regarding the factors affecting selection of destination spots by tourists and what affects them in their selection process. (Tosun *et al.*, 2016: 20)

In this quote it is possible to observe the existence of various sub-decisions in tourism and that the search for belonging may be implied in each one. Thus, the analysis of the opposite of belonging, that which provokes social struggles such as ostracism, becomes essential. After all, knowledge from the neurosciences applied to psychology can contribute to the epistemological advancement of what we know about tourism, as well as to this new paradigm that is currently under construction, neurotourism.

Social Pains: Ostracism

This part of the chapter aims to discuss the social pains caused by ostracism, whilst considering the search for meaning in life and the role played by tourism. As proposed by Donate *et al.* (2017), it is important to explain that ostracism can be understood as a social pain produced when one is excluded or ignored. Nonetheless, physical distancing does not need to be the same as social distancing (Morello & Boggio, 2020); there is a perception that ostracism implies social exclusion, thus the non-belonging to groups. For Myers and Dewall (2017), ostracism is the deliberate social exclusion of individuals or groups. According to Boggio (2021), feelings of social distress can activate the same cerebral areas that are activated by physical pain.

Belonging is one of the strongest bonds in the strengthening of humankind because, generally, it facilitates its own adaptation and survival (Myers & Dewall, 2017). In human beings 'brains have a very well-developed capacity to identify and react to emotions shown by other humans, even with facial microexpressions' (Universidade Presbiteriana Mackenzie, 2021a: 8; our translation). For this reason, as a matter of survival it is primal to be aware of one another and of one another's emotions and feelings. Similarly, there are a series of studies about the biological basis of behaviours that involve empathy, recognition, imitation, and so on. By way of illustration, in neuroscience there are numerous studies on the mirror neuron system in humans, coming from various approaches: neurophysiological, behavioural and brain imaging, emphasising action recognition (Buccino *et al.*, 2004). Largely, the feeling of belonging is owed to a hormone called oxytocin, which is:

> [s]ynthesized in the hypothalamus, the paraventricular nucleus and the supraoptic nucleus. When discharged in determined areas, as the medial prefrontal cortex and amygdala respectively, it boosts confidence, behaviors of cooperation [and] altruism and diminishes defensive behaviors. Furthermore, cooperation and altruistic actions, such as donations, also recruit structures from the reward system. (Universidade Presbiteriana Mackenzie, 2021b: 11; our translation)

Trivers (1971) describes the psychological system based on human reciprocal altruism. Fennell (2006) goes beyond the economic agenda and places the search for the meaning of life at the centre of the debate, since according to Trivers (1971: 45), reciprocal altruism can be understood as (1) helping in times of danger; (2) sharing food; (3) helping the sick, the wounded, or the very young and old; (4) sharing implements; and (5) sharing knowledge. Therefore, by recovering Trivers' (1971) findings on cooperation in the context of stable relationships not necessarily involving kinship, Fennell (2006) uses the sociological theory of reciprocal Altruism to analyse the complexities intertwined within the tourist phenomena. Notably, the author concludes that, because tourism allows limited interactions due to the short time frame, this, among other factors, may prevent long-term cooperation.

Fennell (2006) also reminds us that, according to Trivers (1971), we are driven both by giving and receiving mutual help and by our own interests and competitions. Thus, considering adverse events for tourism, such as the COVID-19 pandemic, it is noteworthy that altruism and ethics factors come out when the theme is the role tourism takes in the search for the meaning of life. Fennell's (2006) study is therefore a relevant invitation to investigate the aspects of altruistic relationships that may occur between the several stakeholders involved in tourism, bringing new clues to understanding the role of tourism in the search for life's meaning.

Actions that express feelings of belonging in tourism can be analysed through various lenses, from the bond between visitors and the 'visited' formed in the cocreation of tourism experiences to the participation in planning and management actions of tourism activities involving the most diverse kinds of stakeholders. In the last few years, much has been discussed in terms of participation in tourism that reflects notions of belonging.

A wider view of the analysis of participation in tourism may suggest that regardless of the side one personifies (visitors/visited, boss/employer, and so on), the objectives, desires, and benefits are always related to the notion of being a part of something, which has to be constructed as a common good for all in terms of sustainable development. It is worth noting that Dardot and Laval (2017) address this in their essay about the notion of 'common' in the 21st century. The authors bring to light the forging of rules and responsibilities as one of the key points of belonging to humanity.

The study of conflicts and subsequent controversies around the idea of belonging related to the dynamics of tourism can offer precise information about human bonds and the search for meaning in life. Therefore, individual and collective responsibilities relative to local development, aiming to achieve sustainability on a global level, can promote win–win relationships between different tourism stakeholders. For instance, it is crucial that the SDGs proposed by the UN (2015) alongside the 2030 Agenda are fundamental guidelines for the launch of any development strategy of touristic destinations on a win–win participation perspective.

In terms of touristic experience, cocreation has assumed a central role in the 21st century through the emergence of so-called creative tourism

(Richards, 2016). As a result, comprehending belonging in tourism may demand the observation of cultural identities (Hall, 2003) in social interactions. The seminal work *Hosts and Guests: The Anthropology of Tourism*, edited by V.L. Smith (1989), is an important lens in this direction. In social terms, the senses and meanings of belonging ascribed to mobilisations, sensibilisations and/or participations in and on tourism in a communitarian framework can shed light on important research paths.

Social pains inflicted by the inverse, that is to say not belonging, not participating, not cocreating, can be observed in the context of social ostracism, a theme not yet widely treated in tourism scientific literature. In tourism, ostracism may occur in various ways – for instance, residents who do not want the presence of tourists may consequently isolate or ignore them. Thus, in extreme circumstances, it is even possible to observe xenophobic acts. Conversely, tourists can also self-isolate in bubbles, ignoring the presence of residents in the destinations they visit. In all cases when there are social pains, the equilibrium required for development in a sustainable way is disturbed.

The overtourism that preceded the health crisis brought on by the COVID-19 pandemic demonstrates the strength that local communities can have to regulate the disorderly growth of tourism. The World Tourism Organization (UNWTO); Centre of Expertise Leisure, Tourism & Hospitality; NHTV Breda University of Applied Sciences; and NHL Stenden University of Applied Sciences (2018) addressed overtourism by creating 11 strategies to manage visitor flows in urban destinations. However, as the COVID-19 pandemic demanded social distancing and the setting up of sanitary barriers in order to contain the virus, both of which ended up inhibiting the global flow of travel, it is noteworthy that future challenges for the resumption of tourism may include controversial and contradictory movements – not only combatting overtourism, which can be triggered by an intense search for several motivations for tourism (e.g. hedonism) after confinement, but also facing the difficulty of recovering the confidence to travel by a given share of potential demand, who have time, financial resources, and motivation but may be stopped by behaviours of aversion and/or a fear of travelling. Therefore, initial social pains caused by ostracism in the pandemic period can have varied consequences, even converting into tourismophobia – that is, a phobia of tourism.

Milano (2018) questions the use of the term tourismophobia and explains that it needs to be analysed beyond the increase in the number of visitors, considering other reasons. The author reinforces that 'in the touristic context it is not only about degrading resources, but also in the breach of the necessary conditions for the tourism activity to be satisfactory to all actors involved' (Milano, 2018: 553; our translation). Nonetheless, it is not only knowledge from anthropology, sociology, and economics that can aid in the examination of this complex tourism phenomenon; advancements in neurosciences open new theoretical and methodological possibilities that have not yet been explored to understand the role that tourism assumes when it comes to the search for meaning in life. For instance, Fennell (2009: 131) shows us that 'pleasure has been a neglected area of study inside and outside tourism'. Thus, exploring the role of tourism in the context of the search for meaning in life includes questions about our motivations when we decide to travel.

Tourism as envisaged in the time and space that it conventionally occupied before the current pandemic, involving the use of transport in order to get to destinations, to accommodation and for sightseeing in general, has given way to other forms of dislocation amidst the COVID-19 pandemic, mostly imaginative, including virtual tourism as a possibility. Virtual tourism is often mediated by a series of technological advancements – augmented reality, artificial intelligence, and so on. During the pandemic, taking a virtual tour of a place located hundreds of kilometres away from one's home became one of the safest ways to exercise the urge to travel. On the other hand, the question may be posed as to whether these practices have the opposite effect, strengthening ostracism and loneliness. Therefore, deepening our consideration of how human motivations connect to notions of belonging can bring powerful elements to the fore for planning and management in aiming to reposition destinations after the end of the pandemic, thereby avoiding the possibility that certain aspects of tourismophobia might gain prominence, endangering sustainable development on a global scale.

Deeper Reflection on Human Motivations

With the propagation of COVID-19 many people had their trips interrupted. Mayer and Coelho (2021) analysed people's memories and emotions surrounding these travelling experiences. Among the most memorable emotions observed in the interviewed subjects – who had their dreams interrupted – were 'strong fear', 'frustration' and 'tension'. It was observed that social pains resulting from social distancing had another dimension amidst the arrival of the pandemic because humanity was required to 'stay home if you can'. Therefore, people were asked to self-isolate in order to promote social distancing and to use masks for self-protection, among other measures that have completely altered routines. In some cases, this contributed to a practically mandatory social ostracism, making it hard to establish the limit between physical and social distancing, although these are not synonymous, as mentioned earlier.

This daily routine brought about by the COVID-19 pandemic has been commonly called the 'new normal', and one of its primary characteristics is the unclear frontier between the real and the virtual – see Lévy (2011) on the definition of *virtuality* – because many work, leisure, and study activities have completely migrated to the virtual universe. It is important to emphasise that this reality comprises the fraction of the population with access to the required capital to allow them to stay at home and make this type of life adaptation. However, discussions surrounding 'normal' and the 'new normal' are diverse and go beyond the continuities and disruptions of the virtual and the real. In tourism, there are relevant criticisms of note about the non-return to 'normal', asserting that it is necessary to search for more equality and that what was identified as 'normal' also needed to be revised, as seen in the argument put forth by Benjamin *et al.* (2020) that COVID-19 has exposed a tourism industry that was already ill and needed a 'reset code'. The authors explain as follows:

> We argue for a call to arms that challenges our purpose as scholars and educators to drive the conversation, to critique, and to take action in planning a post-pandemic tourism. This planning is a multi-dimensional enterprise and requires intervention in the principles and code of ethics of the industry, tourist, and educator. (Benjamin *et al.*, 2020: 478–479)

Ateljevic (2020: 467) is adamant when she claims that '[w]ith or without the global COVID-19 pandemic to promote and envision a meaningful and positive transformation of the planet in general, and tourism specifically, a wake-up call is long overdue'. Hence, the limits between 'normal', the 'new normal' and 'pathological' – regardless of pandemics but still considering them – must be increasingly investigated in tourism in order to contribute to the quest for equity that depends on the sense of us all belonging to the planet (independently of the condition of tourist, resident, etc.). Consequently, the study of human motivations and the quest to find meaning in life during a crucial moment to mankind, such as in the midst of a pandemic, may contribute to the transformations underway and/or to those that are still to come.

Searching within scientific literature is a first step towards comprehending the motivations behind tourism. A search of the Web of Science (WoS) Core Collection on 27 May 2021 for articles in English whose titles contained a combination of the terms 'motivati*' and 'tourism' with the Boolean additional *and*, yielded 387 abstracts, which were later combined into a textual corpus. This corpus was analysed through the employment of the text analysis software Iramuteq (edition 0.7 Alpha 2).

From that first point it was possible to verify the most frequently used terms in this corpus: 'tourist' ($n = 985$) was even more dominant than 'motivation' ($n = 961$). It was also observed that the study of destination and of travel were balanced; that is to say, both the mobile part (travel) and the static part (destination) are equally relevant subjects of research in terms of motivations for travelling. Experiences and the market are also tourism and motivation research targets since these terms (experience: $n = 278$; market: $n = 215$) present respectively high indexes of repetition. Given that the experience of cocreation in tourism is such a key aspect of belonging, and the fact that the market needs an increasingly more inclusive type of planning and management for the achievement of the SDGs for the greater good, it is possible to conclude that studies of motivations in tourism are linked, even if indirectly, to the discussions about belonging as presented in the previous section of this chapter.

In dialogue with Hall (2003) as to the respect of cultural identities, it becomes evident that the cultural aspect is also relevant to the study of motivations in tourism. Hence, social pains must be discussed, considering the search for meaning in life, to determine what motivates the undertaking of tourism in different cultures.

In theoretical terms, it can be observed that the push–pull effect seems to be present because there is an almost parallel frequency of the terms 'pull' ($n = 108$) and 'push' ($n = 104$). Studies that use push–pull are not recent; one of the seminal texts is one by Crompton (1979). However, it is still very much relevant, for instance in the textual corpus that analyses the research by He and Luo (2020) on motivations, satisfaction, and intentions of visiting the Urumqi Silk Road Ski Resort in China based on the push–pull factor.

To illustrate the study on motivations in tourism more profoundly, a cut of equal to or higher than 50 words was made. Considering the category of adverbs as supplementary, an analysis of similarity was made based on graph theory (from mathematics, serving to analyse the relation between objects in a set) in order to illustrate the correlations between terms from the analysed textual corpus (see Salviati, 2017). Then, 'word communities' organised into halos were created, as shown in Figure 2.1. The size of each word is related to the frequency with which it appears in the textual corpus. From this, it is possible to identify relevant themes of interface between tourism and motivations.

The central halo formed by 'motivation' ($n = 961$) correlates directly with the halos formed by 'tourist' ($n = 985$) and 'tourism' ($n = 800$). The motivational relationship of tourists with destinations is therefore quite self-evident. Moreover, sub-halos coming from the central one ('motivation') demonstrate the strength of theoretical and methodological aspects when it comes to the comprehension of the complexity of motivations around tourism, namely the sub-halo for 'model' and 'structural'; another sub-halo for 'approach'; and yet one more for 'design', 'push' and 'pull'.

Based on research evidence, two points emerge: (a) on the side of tourism practice, which has been almost completely paralysed by the COVID-19 pandemic and already has an important demand in terms of developing strategies for the return of travelling after the pandemic, including through a critical view of the 'new normal' and (b) on the side of scientific studies on tourism and motivation, there is still more to be expanded in relation to motivational theories – especially those arising from other areas – in terms of synthesis and discussion. This is the proposition that follows concerning the motivation theories outlined in Table 2.1.

Table 2.1 doesn't exhaust motivation theories but makes it apparent that more than one theory can be used to explain motivations. Moreover, it can

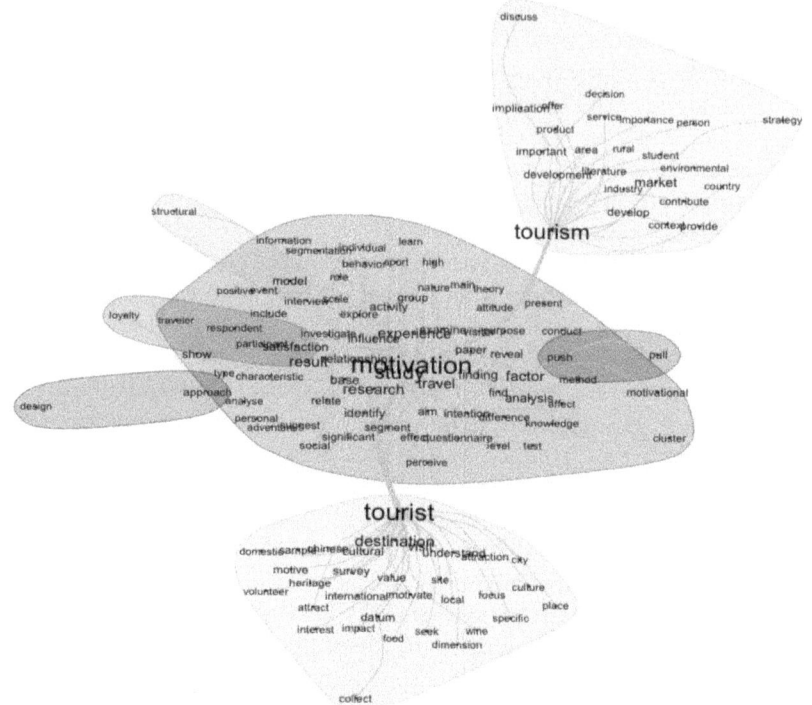

Figure 2.1 The search for meaning in life: the role of motivations in a tourism context
Source: Our elaboration based on consultations of the WoS and Iramuteq software (edition 0.7 Alpha 2)

be observed that these theories can complement each other in certain moments and in others contradict one another. Therefore, the task of researching motivations (be they internal – that is to say, when there is a motivation – or external, when this depends on external rewards; see Ryan & Deci, 2000) is very complex. Feldman (2015: 291) gives a detailed description, for instance, of the fact that 'various psychologists believe that internal impulses proposed by the drive reduction theory work in combination with external incentives of incentive theory for behaviour "attraction" and "impulse", respectively'. Feldman (2015: 291) also highlights other approaches to the study of motivations, such as the cognitive approach, which in short is composed of theories 'that suggest that motivation is a product of thoughts, expectations and people's objectives – their cognitions'. Also relevant is self-determination theory, in which intrinsic and external motivation are equivalent to the same context – that is to say, attitude plus objectives motivate action (see Ryan & Deci, 2000; Ryan & Deci, 2017).

According to instinct theory, some behaviours have been identified in accordance with fixed patterns (which are learned), but this can be questioned. With drive reduction theory it is possible to inquire how physiological

Table 2.1 Motivation theories

Theories	Description
Instinct theory (now also called evolutionist perspectives)	This relates to genetically predisposed behaviours, fixed patterns that are not learned. There is an observable repetition of patterns across members of the same species. The expansion of evolutionary theory by Charles Darwin in the 20th century backed up this theory. However, instinct theory presents a series of considerations.
Drive reduction theory	In this theory it is clear that physiological necessity unleashes a reaction that propels the organism forward (or drives it) to reduce such need and create an alert. Positive and negative incentives can also influence this. Therefore, individual learning affects motivation. This theory connects to the necessity of homeostasis conservation, namely, the dynamic balance of an organism.
Optimal arousal theory	This theory interacts with the notion of engagement from body homeostasis. In other words, it is not only the reduction of impulse excitement but the search for optimal levels of excitement and arousal. The optimal level of arousal allows for better adaptation to the environment. Relevant in this connection is Yerkes–Dodson law – Yerkes and Dodson (1908) observes that the optimal quality of performance happens when the physiological alert level is moderate.
Hierarchy of needs theory	Abraham Maslow developed a theory in the 1970s that came to be known for its pyramid design. In this pyramid, the priority order is in an upward direction considering physiological needs, safety needs, belonging and love needs, esteem needs, and self-fulfilment. Therefore, for Maslow, some needs are more basic and urgent than others.

Source: Our elaboration based on consultations of Myers and Dewall (2017).

innate needs and incentives from the environment influence impulses and behaviour. However, choice motivations are not only the product of instincts and/or the search for homeostatic balance that these theories highlight.

According to Myers and Dewall (2017: 347), '[H]uman motivation doesn't try to eliminate arousal, but to find the optimal levels'. For this reason, the Yerkes–Dodson law about performance and arousal levels is an important indicator for understanding motivations. In turn, Maslow's proposal can contribute to the study of priorities in the decision-making process in terms of what motivates us. Nevertheless, this theory also presents its own controversies; for example, in some cases, couldn't the needs for belonging and esteem come before the need for safety? How many times have people done risky things in the hope of being accepted? According to Feldman (2015: 292), '[S]ome people look for high levels of arousal, whilst others are less preoccupied'. Myers and Dewall (2017: 370) posit that motivation for accomplishments can be understood as 'a desire for significant realization; for mastering skills and ideals; for control; and for reaching a high standard'. The authors reach the heart of the question of belonging, connecting it to motivations:

> The biological perspective on motivation – the idea that physiological necessities lead us to fulfill our other needs – provides a partial

explanation of what energises and directs one's behaviour. Hunger and the need to belong have social and biological components. Moreover, there are reasons which seem to have obvious low value for survival. (Myers & Dewall, 2017: 370; our translation)

Thus, based on these theories, it becomes clear that there are biological and non-biological perspectives through which to understand motivations. This point is exactly where the alliance between neurosciences and psychology can contribute to the study of tourism, notably for the role of tourism in the search for meaning in life, since it is all about understanding motives (biological and/or psychological) that relate to survival, even if these are not obvious. Ergo, it is also in this specific situation that consideration of social pains, such as ostracism and the sense of belonging, one of its opposites, becomes essential as an element in planning and management in tourism, since one of the main deliverables of tourism is precisely otherness, translated by the quest to get to know the other. In this sense, motivational research can be quite useful to studies of tourism.

It [w]as developed depending on a change that happened over the years in the supply–demand relationship, in view of the fact that individuals started to choose products from among so many others that were available. For this reason, companies started developing products that were in accordance with consumers' desires. (Pires, 2018: 142; our translation)

Another point to be made concerns the reward system, though this has not yet been studied exhaustively in tourism studies based on neurosciences. Rewards involve feelings of pleasure, the component of incentive, namely, what motivates us, and the learning that rewards drive. As mentioned above, the COVID-19 pandemic is challenging for tourism because of social distancing and the sanitary barriers that were imposed on the tourism and travel sector. These forced a rethinking of the meaning of common good and a realisation of the need to cocreate an increasingly healthier future for this industry, which includes paying attention to the motivations of local community members and tourists. Therefore, both belonging and social pains have become an essential subject for analysis, through multidisciplinary and interdisciplinary theoretical and methodological lenses, and the importance of motivation has been recognised in terms of decisions and behaviours in tourism.

Conclusion

The construction of a new theoretical and methodological basis for viewing the tourism phenomenon – at the centre of the search for meaning in life – can contribute to the rise of a new epistemological perspective for its treatment. On one hand, elements related to belonging, hedonism, and happiness are always undergoing transformations (over time – in terms of the history of the neurosciences addressed by this article); on the other hand, elements related to social pains (such as ostracism) need to be more profoundly understood.

That being said, we return to the heart of one of the biggest questions in tourism studies: What role does this complex phenomenon play in the search for meaning in life? Although this might be an individual response, it is the study of motivations through various disciplinary lenses, including alliances between neurosciences and applied psychology, among other fields, that will be able to deliver what is yet to be discovered. Therefore, going back to Kandel (2020), it is important to highlight that advancements in our understanding of the mind's biology are directly linked to a new type of humanism, although an alliance between the sciences (natural and social) is still needed. This question – which at first glance resembles a metonym but is, in fact, a metaphor – forces a hard exercise that has continuously been postponed, demanding not only financial resources for basic and applied research but also the orchestration of teams who will be able to apply multi and interdisciplinary perspectives, or maybe even postdisciplinary ones.

Finally, the theoretical-conceptual and practical implications of this chapter can be synthesised as follows: (1) deepening the understanding of neurosciences history will offer support to that which has been developed as neurotourism; (2) the systematic approach of neuroscience theories and methods applied in applied psychology has great potential for studies in tourism, delivering possibilities for the forging of new paradigms; (3) the investigation of social pains, notably ostracism, as relevant elements for the episteme and for the planning and management of tourism can elevate tourism studies to another level, as well as enabling the development of practices to deal with tourism's existing problems; (4) the refinement of instruments and the design of motivational research that goes beyond what is already known in order to include neuroscientific methods and techniques, including deepening those about social interactions (like hyperscanning techniques); (5) the quest for furthering what is known in the neurosciences about the reward system in a tourism motivational context is an important pathway for the development of that which motivates tourism in general. All these implications must be accompanied by rigorous research ethics, even more so when neurotourism is considered as a new paradigm and/or a new discipline.

References

Ateljevic, I. (2020) Transforming the (tourism) world for good and (re)generating the potential 'new normal'. *Tourism Geographies* 22 (3), 467–475. https://doi.org/10.1080/14616688.2020.1759134.

Azmitia, E.C. (2015) Cajal e a plasticidade cerebral: Ideias relevantes para conceitos emergentes da mente. In F.R.M. Ferreira and M.I. Nogueira (eds) *História da filosofia da neurociência* (pp. 99–119). São Paulo: LiberArs.

Benjamin, S., Dillette, A. and Alderman, D.H. (2020) 'We can't return to normal': Committing to tourism equity in the post-pandemic age. *Tourism Geographies* 22 (3), 476–483. https://doi.org/10.1080/14616688.2020.1759130.

Boggio, P.S. (2021) *Trilha 3 Pertencimento e Motivação no Trabalho*. Neurociência e Psicologia Aplicada. Trabalho, motivação e stress. Mackenzie. Professor Paulo Boggio.

Buccino, G., Binkofski, F. and Riggio, L. (2004) The mirror neuron system and action recognition. *Brain Lang* 89(2), 370–376. https://doi.org/10.1016/S0093-934X(03)00356-0.

Carlos, J.A. de. (2015) Santiago Ramón y Cajal: Uma biografía científica. In F.R.M. Ferreira and M.I. Nogueira (eds) *História da filosofia da neurociência* (pp. 11–47). São Paulo: LiberArs.

Crompton, J.L. (1979) Motivations for pleasure vacation. *Annals of Tourism Research* 6, 408–424.

Dardot, P. and Laval, C. (2017) *Comum: Ensaio sobre a revolução no século XXI*. São Paulo: Boitempo.

Donate, A.P.G., Marques, L.M., Lapenta, O.M., Asthana, M.K., Amodio, D. and Boggio, P.S. (2017) Ostracism via virtual chat room—Effects on basic needs, anger and pain. *PLoS ONE* 12 (9), e0184215. https://doi.org/10.1371/journal.pone.0184215.

Feldman, R.S. (2015) *Introdução à psicologia [electronic resource]* (10th edn). Translated by D. Bueno and S.M.M. da Rosa. Technical review by M.L.T. Nunes. Porto Alegre: AMGH.

Fennell, D.A. (2006) Evolution in tourism: The theory of reciprocal altruism and tourist-host interactions. *Current Issues in Tourism* 9 (2), 105–124. https://doi.org/10.1080/13683500.2013.850064.

Fennell, D.A. (2009) The nature of pleasure in pleasure travel. *Tourism Recreation Research* 34 (2), 123–134.

Giudici, E., Dettori, A. and Caboni, F. (2017) *Neurotourism: Futuristic Perspective or Today's Reality?* 20th Excellence in Services International Conference Verona. University of Verona, Verona, 7–8 September. Conference proceedings.

Gurgel, I. (2017) Sobre a importância da História das Ciências. *Journal da USP*, 1 November. See https://jornal.usp.br/artigos/sobre-a-importancia-da-historia-das-ciencias (accessed May 2021).

Hall, S. (2003) *A identidade cultural na pós-modernidade* (7th edn). Translated by T.T. da Silva and G.L. Louro. Rio de Janeiro: DP&A.

He, X. and Luo, J.M. (2020) Relationship among travel motivation, satisfaction and revisit intention of skiers: A case study on the tourists of Urumqi Silk Road Ski Resort. *Administrative Sciences* 10 (3), 56. https://doi.org/10.3390/admsci10030056.

Kandel, E.R. (2020) *Mentes diferentes: O que os cérebros incomuns revelam sobre nós*. Barueri: Manole.

Lévy, P. (2011) *O que é virtual?* Translated by Paulo Neves. São Paulo: Editora 34.

Lohmann, G. and Panosso Netto, A. (2008) *Teoria do turismo: Conceitos, modelos e sistemas*. São Paulo: Aleph.

Mayseless, N., Hawthorne, G. and Reiss, A.L. (2019) Real-life creative problem solving in teams: fNIRS based hyperscanning study. *Neuroimage* 203, s116161. https://doi.org/10.1016/j.neuroimage.2019.116161.

Mayer, V.F. and Coelho, M.F. (2021) Sonhos interrompidos: Memórias e emoções de experiências de viagem durante a propagação do COVID-19. *Revista Brasileira de Pesquisa em Turismo* 15 (1), article ID 2192. https://doi.org/10.7784/rbtur.v15i1.2192.

Mella, F.A.A. (2001) *Dos Sumérios a Babel*. São Paulo: Hemus.

Milano, C. (2018) *Overtourism*, malestar social y turismofobia: Un debate controvertido. *Pasos Revista de Turismo y Patrimonio Cultural* 16 (3), 551–564.

Moraes, A. (2009) *Parahyba Quartim de – O Livro do cérebro*. Volume 1. São Paulo: Editora Duetto.

Morello, L.Y. and Boggio, P. (2020) Isolamento físico não precisa ser isolamento social. *Revista Questão de Ciência*, 11 April. See www.revistaquestaodeciencia.com.br/artigo/2020/04/11/isolamento-fisico-nao-precisa-ser-isolamento-social (accessed May 2021).

Myers, D.G. and Dewall, C.N. (2017) *Psicologia* (11th edn). LTC.
Panosso Netto, A. (2010) *O que é turismo*. São Paulo: Brasiliense.
Pires, L.R. (2018) Aprendizagem e motivação: Necessidades, desejos, emoção, ação e instinto. In L.R. Pires *et al.* (eds) *Psicologia [recurso eletrônico]*. Technical review by C. Capaverde and A. Canaparro da Silva. Porto Alegre: SAGAH.
Prideaux, B., Thompson, M. and Pabel, A. (2020) Lessons from COVID-19 can prepare global tourism for the economic transformation needed to combat climate change. *Tourism Geographies* 22 (3), 667–678.
Richards, G. (2016) The challenge of creative tourism. *Ethnologies* 38 (1–2), 31–45. https://doi.org/10.7202/1041585ar.
Rocha, E.T., Alves, T.CTF., Garrido, G.E.J., Buchpiguel, C.A., Nitrini, R. and Filho, G.B. (2001) Novas técnicas de neuroimagem em psiquiatria. *Revista Brasileira Psiquiatria* 23 (sup. 1), 58–60.
Rooney, A. (2018) *A história da neurociência: Como desvendar os mistérios do cérebro*. São Paulo: M. Books do Brasil Editora.
Ryan, R.M. and Deci, E.L. (2000) Self-determination theory and the facilitation of intrinsic motivation, social development, and well-being. *American Psychologist* 55(1), 68–78 . https://doi.org/10.1037/0003-066X.55.1.68.
Ryan, R.M. and Deci, E.L. (2017) *Self-Determination Theory: Basic Psychological Needs in Motivation, Development, and Wellness*. New York: The Guilford Press. https://doi.org/10.1521/978.14625/28806.
Salviati, M.E. (2017) *Manual do Aplicativo Iramuteq*. Planaltina. See www.iramuteq.org/documentation/fichiers/manual-do-aplicativo-iramuteq-par-maria-elisabeth-salviati (accessed May 2021).
Scott, N. (2020) Cognitive psychology and tourism – surfing the 'cognitive wave': A perspective article. *Tourism Review* 75 (1), 49–51. https://doi.org/10.1108/TR-06-2019-0217.
Sigala, M. (2020) Tourism and COVID-19: Impacts and implications for advancing and resetting industry and research. *Journal of Business Research* 117, 312–321. https://doi.org/10.1016/j.jbusres.2020.06.015.
Smith, V.L. (ed.) (1989) *Hosts and Guests: The Anthropology of Tourism* (2nd edn). Philadelphia: University of Pennsylvania Press.
Teixeira, J. de F. (2015) Imagens do cérebro. In F.R.M. Ferreira *et al.* (eds) *História da filosofia da neurociência* (pp. 173–182). São Paulo: LiberArs.
Tosun, C., Ozdemir, A.S. and Cubuk, F. (2016) Usage of neuro-tourism methods in hotel preferences of the consumers. In *The 2016 WEI International Academic Conference Proceedings*. The West East Institute, Boston. Conference proceedings.
Trivers, R.L. (1971) The evolution of reciprocal altruism. *The Quarterly Review of Biology* 46 (1), 35–57.
United Nations (2015) *Transforming our World: The 2030 Agenda for Sustainable Development*. New York: United Nations.
Universidade Presbiteriana Mackenzie (2021a) Neurociência e Psicologia Aplicada. Trabalho, Motivação e Stress. Módulo 3. Trilha 3.
Universidade Presbiteriana Mackenzie (2021b) Neurociência e Psicologia Aplicada. Trabalho, Motivação e Stress. Módulo 3. Trilha 4.
World Tourism Organization (UNWTO); Centre of Expertise Leisure, Tourism & Hospitality; NHTV Breda University of Applied Sciences; and NHL Stenden University of Applied Sciences (eds) (2018) *'Overtourism'? Understanding and Managing Urban Tourism Growth beyond Perceptions*. Volume 2: *Case Studies*. Madrid: UNWTO. https://doi.org/10.18111/9789284420629.
Yerkes, R.M. and Dodson, J.D. (1908) The relation of strength of stimulus to rapidity of habit-formation. *Journal of Comparative Neurology and Psychology* 18, 459–482.

3 *Eat Pray Love* or *Total Recall*? Mindfulness and Tourism

Gianna Moscardo

Eat Pray Love and *Total Recall* are two popular films that present very different visions of tourism. In *Total Recall* (1990), a science fiction story, the future is a dark and bleak place where most of the population are poor, often suffering from genetic mutations, and confined to dangerous jobs and unpleasant living and working conditions. In this bleak and not altogether implausible future, most ordinary people can only travel through experiences implanted directly into their brains. The story explores what happens when such an implant goes wrong, sending the story's protagonist on a desperate journey to find his true self. *Eat Pray Love* (2010) is a romance story set in the contemporary (pre-COVID) world and based on real world events. The protagonist of this story runs away from her everyday life in the wake of failed relationships, searching for happiness and some sense of her true self. In Italy she learns to enjoy food, in India she learns to pray, and in Indonesia she finds true love.

Both our protagonists travel in an attempt to find their true selves. In *Total Recall* the traveller chooses to be a hero defying and then defeating the corporate overlords, freeing the poor from the negative environmental consequences of unchecked development. In *Eat Pray Love*, it is not as clear that the traveller finds her true self, but she does find happiness and supposedly true love. Both examples are stories about travel as transformation, albeit in very different ways. *Eat Pray Love* is light and beautiful, an advertisement for the three destinations visited, offering rich, young, affluent white people the chance to find happiness by travelling somewhere else. This is a common theme in academic discussion of how tourism can be transformative. *Total Recall* is dark and violent and acts almost as a travel advisory warning for would-be tourists to stay at home. Arguably this is also a theme, although not such a common one, in academic discussions of the challenges of tourism and sustainability.

If our *Eat Pray Love* heroine were to embark on her adventure today it seems likely she would spend some time engaging in mindfulness meditation, as this is the fashionable, vaguely Eastern, and alternative addition to yoga practices for affluent white tourists in wellness centres across the globe. This chapter will focus on the concept of mindfulness and how it has been, and could be, used in tourism to enhance well-being. It will begin by describing what mindfulness is, introducing the different approaches to mindfulness and where they converge and diverge. After a critical overview of the use of mindfulness in tourism research to date, the chapter will consider whose mindfulness and where, how, and why mindfulness matters for tourism. The chapter will identify the various ways mindfulness could assist in improving the positive contributions that tourism can make to both tourists and hosts. Finally, the chapter will return to the two stories described at the start of the chapter, noting that in good stories things do not always go as expected.

Mindfulness: One Concept, Multiple Pathways

Before discussing the links between mindfulness and well-being in tourism it is important to have a clear idea of what mindfulness is. There has been an almost exponential rise in discussions of mindfulness in research and practice within and beyond psychology in the last decade. Marks and Cowan (2017) report that a search of PsycINFO (the primary academic research database for psychology, which should be used much more by tourism researchers borrowing from this discipline) in 2005, revealed 41 papers with mindfulness in the title. A similar search conducted by this chapter's author for 2020 revealed 739 such papers, reflecting the widespread adoption of mindfulness as a therapeutic technique in psychological counselling, in addition to its commodification as the latest wellness trend amongst affluent, predominantly white consumers in the countries of the global north. Unfortunately, while mindfulness is clearly defined in psychology, enthusiasm for the mindfulness–wellness trend has resulted in some non-psychologists, including tourism academics, skipping some of the key scholarly steps, confusing the use of mindfulness terms and definitions and repeating popular claims about mindfulness outcomes without proper examination of the available research evidence. Tisdell and Riley (2019) note that both popular and academic discussions of mindfulness often confuse mindfulness with meditation, which in turn is also sometimes called meditative mindfulness. This is further confused by a lack of awareness and/or acknowledgement of the different approaches to mindfulness that exist within psychology (Marks & Cowan, 2017).

It is important to distinguish between mindfulness and meditation and this chapter will argue that we can identify several distinctive mindfulness approaches. Mindfulness as a cognitive state is defined in a common way regardless of which of the various approaches apply. While the following section discusses mindfulness definitions in more detail, mindfulness is described in all the different approaches as a state of active cognitive engagement characterised by focused attention and awareness of the present moment that allows for development of new and more appropriate responses to external events and conditions. Meditation is a general practice that has long been claimed to enhance personal well-being and diverse meditation techniques can be found both across and within all major religions (Tisdell & Riley, 2019). While many Western devotees have adopted meditation practices from Hindu, Buddhist and Taoist traditions, each of which has multiple meditation techniques, examples of similar meditative practices exist in Christian, Muslim and other religious practices (Oman, 2020). Mindfulness-based meditation is often incorrectly described by non-psychologists drawing on the work of Jon Kabat-Zinn (2005) as derived from Buddhism (cf. Chen *et al.*, 2017). The use of meditation techniques as therapy in general, and as a pathway to mindfulness specifically, predates Kabat-Zinn, and the range of meditation techniques used extends a great deal further than his very specific options. While this approach in general argues that mindfulness is achieved through meditation, mindfulness itself is still considered a separate and distinctive state. Put simply, not all meditation is about mindfulness and mindfulness is not meditation. These distinctions matter in academic discussions because they connect concepts to different assumptions, values, desired outcomes, and processes. Confusing them leads to poor methodology and confused and incorrect conclusions. These problems exist in abundance in applied areas such as tourism (cf. Lengyel, 2020).

Regardless of the different development histories, all psychologists share very common definitions of mindfulness as a cognitive state. Psychologists recognise multiple possible pathways to mindfulness, usually derived from two complementary, rather than conflicting, traditions: the therapeutic use of meditation to help patients manage pain, stress and anxiety, which shall be referred to as meditation-based mindfulness (MBM), and the sociocognitive dual-processing tradition now often referred to as Langerian mindfulness (LM) because of the dominance of work by Ellen Langer, her students, and her colleagues (Fatemi, 2016b). Beyond psychology there are other mindfulness movements and so this chapter argues for at least four major approaches to mindfulness: LM, MBM, Buddhist practices linked to the concepts of *Sati*, and what is increasingly being referred to as McMindfulness (Hyland, 2017).

McMindfulness or commodified Buddhist-linked meditation-based mindfulness (BMM) mostly exists outside of psychology. McMindfulness has been adopted by some counselling psychologists in response to consumer demand, and with great enthusiasm by researchers in applied areas such as sustainability, education, and tourism. It refers to the commodification of varied meditation practices and the selective cultural appropriation of some elements of Buddhism. Farias and Wikholm (2016: 329) argue that while mindfulness 'is a powerful social phenomenon, [it is] rooted in our [Western] culture's desire for quick fixes and its attraction to spiritual ideas divested of supernatural elements'. They further argue that this is 'a misrepresentation of the place and value of meditation in the Buddhist tradition, including its depiction as a purely rational method of self-exploration, which would feel alien to countless generations of Buddhists'. Not surprisingly, some Buddhists take offence at the appropriation of a single element of their religion by those with little or no understanding of, or commitment to, the entire set of spiritual beliefs embedded in Buddhism (Gethin, 2011; Poceski, 2020). It is hard to imagine scholars being so casual and superficial in their adoption of similar concepts from other spiritual belief systems and yet none of the tourism scholars arguing for this Buddhist-flavoured MBM seem aware of these issues. Three groups of tourism researchers directly link mindfulness to Buddhist practice with none of them acknowledging any critiques of such a practice and none of them describing how the removal of meditation from its broader spiritual context is likely to achieve any of the outcomes they claim (Chen et al., 2017; Lengyel, 2020; Stankov et al., 2020a, 2020b). It is also worth noting that similar concepts to *Sati* exist in Christianity (Trammel, 2017), Islam (Thomas et al., 2017), Judaism (Niculescu, 2020), and Confucianism (Tan, 2021). This has also been ignored in almost all tourism papers on meditation and mindfulness. In many tourism papers the link to Buddhism seems to be used as a vague argument to support the claim that meditation is a legitimate way to reconsider how we relate to the world around us (cf. Lengyel, 2015). Buddhist discussions of meditation and mindfulness demonstrate considerable debate about whether the concept of *Sati* can actually be interpreted as mindfulness as this label is used in secular Western contexts (Gethin, 2011; Nilsson & Kazemi, 2016; Trammel, 2017). There is also a very clear argument that taking specific meditative practices out of Buddhism is not a pathway to *Sati*/mindfulness and that, without the additional ethical considerations that come from embracing the entirety of Buddhism, the meditation element on its own is unlikely to change anyone or anything for the better (Farias & Wikholm, 2016; Hyland, 2017).

A further major criticism of McMindfulness is its focus on an individual's acceptance of their current situation and the assumption that constant happiness is a desirable state of being (Hickey, 2010). To quote from another story, this time a novel about the psychological damage passed down through generations of a family desperately seeking a shortcut to constant happiness, *The Seed Collectors* by Scarlett Thomas (2015: 155–156), '[M]indfulness . . . seems intended to turn you into a docile animal that stands in its field all day, never complaining and never smashing down fences'. Several sociologists have begun to explore how the adoption of McMindfulness suits neoliberal and capitalist agendas by discouraging questioning of systemic issues that underpin social injustice, economic inequality and environmental degradation (cf. Cook, 2016; Purser, 2018; Scherer & Waistell, 2017).

Mindfulness Definitions

Even a cursory examination of mindfulness definitions from LM and MBM demonstrates their similarity. In LM, for example, mindfulness is defined as 'a flexible state of mind in which we are actively engaged in the present, noticing new things and sensitive to context' (Langer, 2000: 220), or as a cognitive state with the following qualities: '(1) creation of new categories; (2) openness to new information; and (3) awareness of more than one perspective' (Langer, 2014: 64). In MBM, Kabat-Zinn's (2014: 342) oft-quoted definition is that mindfulness is a state of 'moment-to-moment, non-judgemental awareness', while Siegel (2007: 5) defines it as 'waking up from a life on automatic, and being sensitive to novelty in our everyday experiences'. Contrary to many applied research discussions of mindfulness, mindfulness definitions are remarkably similar across LM and MBM. The argument that mindfulness is defined very differently by Langer versus Kabat-Zinn was initially proposed in tourism by Chen and colleagues in tourism in 2017 and then subsequently accepted in further discussion of MBM in tourism (cf. Lengyel, 2020). Unfortunately, even a casual reading of contemporary literature in the LM and MBM traditions would have revealed that the elements claimed by Chen *et al.* (2017) as different to LM are in fact central to definitions of LM and that the elements claimed to be specific to LM are also typically included in definitions of MBM.

All mindfulness definitions across all the traditions or approaches share the features of awareness of the present, directed attention, avoidance of automatic responses, ability to create and use novel/new responses, and cognitive engagement. This is consistent with Nilsson and Kazemi's (2016) systematic and thematic review of definitions of mindfulness. This review suggested, after excluding Buddhist discussions of mindfulness, that there are four common elements of mindfulness definitions that exist across all the psychological discussions. These can be used in a working definition suitable for both psychology approaches: mindfulness is a state of active cognitive engagement characterised by focused attention and awareness of the present moment that allows for development of new and appropriate responses to external events and conditions. Mindfulness is therefore a foundational element of well-being. This critical examination suggests that the different approaches share the common goal of encouraging a mindful state of being and share a common conception of what this state is. The critical difference between LM and MBM is in their perspective on the importance of meditation in encouraging mindfulness.

Overview of Langerian Mindfulness

LM is explicitly based on the concept of dual processing in social cognition. Dual processing is a fundamental concept in psychology (Evans,

2018) that states that in any given situation a combination of personal and situational variables encourages individuals to be in one of two possible cognitive states:

(1) shallow, fast, or heuristic thinking that is characterised by reliance on stereotypes or established routines, and limited attention to and processing of new information (Evans, 2018), referred to by Langer (2014) as mindlessness; or
(2) deep, slow, or systematic thinking that is characterised by cognitive engagement, attention to and detailed processing of information available in the setting (Evans, 2018), referred to by Langer (2014) as mindfulness.

Consistent with dual-processing assumptions, mindfulness results from a combination of, and interaction between, personal and situational variables. Langer (2014) argues that changes in situational conditions can encourage mindfulness, offering guidance for educators, communicators, and managers in designing places and programmes to encourage and support the mindfulness of users. She also argues that changes within the individual can be supported by a variety of techniques, including creativity exercises and training in meta-cognitive skills and meditation (Langer, 2014). Many of the situational variables and interventions can be seen as mechanisms for disrupting automatic, mindless routines, or conditions that require and then support continued mindfulness. These interventions can be seen as covert in that they do not require the targets of the interventions to be aware of the interventions as anything other than ordinary elements of the situation. These covert situational interventions can be contrasted with more overt and personal approaches in which the target individuals are explicitly informed about, or even actively seek to engage in, exercises to develop their ability to consciously choose to be mindful and avoid automatic responses regardless of the setting and its cues. Table 3.1 lists some of the most common interventions in each of these categories. It should be noted that some of the meta-cognitive skills interventions in the overt category, such as creativity exercises, guided multisensory attention-focusing activities, and play, can also be incorporated into tourist experiences as rewarding elements without any labelling of them as mindfulness pathways.

In terms of outcomes, LM, resulting from either situational or personal interventions, has been found to support improved memory, greater learning, enhanced creativity, increased sense of control and self-efficacy, greater likelihood of flow experiences, positive emotions in general, less fatalistic views, changes in attitude, greater intention behaviour consistency, greater awareness of personal values and increased likelihood of matching actions and decisions to personal values, and more effective decision-making (see various chapters in the edited volumes by Baltzell, 2016; Ie et al., 2014a; Fatemi, 2016a; Fischer et al., 2017; Helm &

Table 3.1 Situational and personal interventions to encourage mindfulness

Situational/Covert	Personal/Overt
Giving people choices and control over the direction of the experience and/or responsibility for some section of the experience	Creativity exercises that require people to challenge existing assumptions and connections
Providing variety in presentation of information and on all dimensions of experiences including level of physical challenge, types of cognitive activities, and types of physical settings	Specifically questioning and avoiding the use of labels
	Games requiring the creation of new categories
Variety in themes and/or topics to allow individuals to build personal connections	Adaption of children's play, especially imaginative play such as finishing a story or creating a story from an image
Creating multisensory settings	Learning to reframe processes and problems
Encouraging or requiring active participation on some aspect of the setting or the experiences rather than being a passive audience	Getting people to draw their own conclusions from the presentation of different information
	Learning to identify and avoid faulty comparisons and to create new analogies/metaphors
Presenting a range of different perspectives on the same object or topic	Conditional learning exercises, for example things are presented as 'could be' rather than 'as'
Immersion in an experience/setting	
Encouraging collaborative problem-solving and responsibility for outcomes	Using language descriptively rather than judgementally
Encouraging people to find a personal connection to the experience, setting, or topic	Mindful attention to setting details/attention training/motivated perception
Using the unexpected or novel to break routines	Arguing both for and against a position

Sources: Balcetis *et al.*, 2014; Herbert, 2014; Kang *et al.*, 2014; Kanter & Fox, 2016; Langer, 2014; Moldoveanu, 2016; Moscardo, 1999, 2017a; Niedderer, 2014.

Subramaniam, 2019; Ivtzan & Hart, 2016; Pagnini & Phillips, 2015; Wamsler & Brink, 2018). However, it is important to recognise that mindfulness is a necessary but not sufficient condition for these outcomes. Being mindful, while a necessary condition for learning, will not necessarily result in learning if the material presented is incomprehensible or not logically connected or well supported by evidence. Being mindful, for example, may make people aware of how poorly they are being treated and this may lead to dissatisfaction and unhappiness. Similarly, mindful individuals may become more aware of their values, but if those values are oriented towards social status and conspicuous consumption then they may pursue unsustainable behaviours even more efficiently than before. In other words, being mindful can make you question the systems around you in ways that are unpredictable, and mindfulness does not guarantee that you will be a moral person or behave in an ethical fashion. While there has been some debate about specific details of LM, such as whether or not mindfulness is both a trait and a state, whether it is a cognitive state or metacognition, how it relates to other concepts such as flow and field dependence, and how mindfulness can be measured, there exist no major critiques of the theoretical framework setting out the processes and conditions that contribute to mindfulness, or of the work linking mindfulness to various well-being related outcomes. Despite the evidence supporting the proposed processes that encourage mindfulness and linking mindfulness to improved outcomes, it is important to reiterate the point made earlier, that while mindfulness may be a necessary condition for the positive outcomes, it is not in itself sufficient to achieve these outcomes. The additional steps needed to link mindfulness at an individual level to improved outcomes from tourism as a socioeconomic phenomenon will be discussed in the final section of the chapter.

The four main differences between LM and MBM are (1) the importance of meditation as a pathway to mindfulness, (2) the outcomes of interest (Ivtzan & Hart, 2016), (3) the amount of effort required from individuals and (4) the proposal that mindfulness can be a trait as well as a state. LM advocates recognise that meditation may support individuals in improving their mindfulness but argue that meditation is not the sole pathway to mindfulness (Pagnini & Phillips, 2015). MBM advocates instead argue that meditation is essential to achieve mindfulness. It should be noted, however, that the majority of MBM programmes, including those championed by Kabat-Zinn, incorporate meta-cognitive skills training components derived from LM. LM seeks to encourage mindfulness to support overall well-being, assisting people to make better decisions, improve their learning and performance at work and at leisure, be more creative, and be safer and healthier. MBM is focused on managing specific medical conditions such as chronic pain and anxiety- and stress-related disorders, so is much closer to medical practice and alternative medical therapies – although the wider adoption of MBM as a general wellness technique is broadening this focus to well-being in general. Given the central role of meditation, MBM interventions are overt and require a considerable investment of time and effort on the part of target individuals, with most MBM programmes lasting 8 to 10 sessions either intensively over a 10- to 14-day period or over at least 8 weeks with an ongoing commitment to daily meditation and other cognitive therapy practices. Given the current fashionable enthusiasm for mindfulness there has been considerable growth in commodified versions of MBM programmes, especially mobile apps, supposedly designed to help people more easily engage in meditation. There is, however, no evidence that any of these are effective, and most psychologists who advocate MBM are sceptical of the value of these clearly commercial exercises. LM interventions can be either overt and covert and even when overt can be more easily incorporated into daily life. Some descriptions of mindfulness in the MBM approach seem to suggest that meditative practice is meant to encourage participants to be more mindful in general in their everyday lives. LM also argues that the development and practicing of various meta-cognitive skills can encourage more mindful approaches to life in general. In LM there is, however, also recognition that mindfulness can be a trait – that is, some people are more mindful in general, and this has been referred to as dispositional mindfulness (Siegling & Petrides, 2014).

Meditation-based Mindfulness

The core arguments supporting the use of meditation as a therapeutic tool in general is that it allows individuals to pause and remove themselves from usual patterns of thinking that can be destructive and stressful and gives them the opportunity to observe and learn to recognise and control how their body reacts physically to stress and negative patterns of thinking and acting (Hölzel et al., 2011). MBM advocates argue that the meditation pause also allows people to re-evaluate how they respond to their lives in general and therefore encourages them to be able to switch from mindless to more mindful routines (Hölzel et al., 2011; Siegel, 2007; Tisdell & Riley, 2019). Like LM interventions, meditation is a personal tool that can be used to disrupt mindlessness, opening the possibility of taking a more mindful stance to life general. As noted previously, most MBM interventions include elements that encourage individuals to also learn and practice meta-cognitive skills (Siegel, 2007).

While there is some evidence to support the beneficial outcomes of MBM, numerous systematic and metanalytical reviews of research evaluating MBM interventions conducted by both psychologists and researchers in applied areas such as education and sustainability action have consistently concluded that the research to support MBM interventions, including both self-report and neurological brain imaging studies, is plagued by major methodological weaknesses (Chiesa & Serretti, 2009; Farias & Wikholm, 2016; Geiger et al., 2019; Hickey, 2010; Jamieson & Tuckey, 2017; Lee, 2019; Thiermann & Sheate, 2020; Van Dam et al., 2018). The identified methodological issues include small sample sizes, homogeneity of samples, unclear definitions of mindfulness and outcomes being measured, inadequate controls, inability to distinguish the effects of meditation from the effects of other intervention elements, self-selection into interventions, issues in interpretation of brain images for neurological research, and problems with reactivity and history effects in experiments. Furthermore, even when research that is not compromised by these methodological weaknesses is examined, the evidence suggests only moderate short-term benefits for some participants, with almost no evidence at all supporting medium and long-term benefits and no evidence that MBM interventions are more effective than drugs, physical exercise, relaxation techniques without meditation, cognitive therapies without meditation, supportive social interaction, or traditional counselling techniques (Chiesa

& Serretti, 2009; Farias & Wikholm, 2016; Geiger *et al.*, 2019; Hickey, 2010; Jamieson & Tuckey, 2017; Saeed *et al.*, 2010; Thiermann & Sheate, 2020; Van Dam *et al.*, 2018). Finally, there is evidence that for some people meditation can be very harmful by bringing back traumatic incidents, challenging their sense of self in a negative way, and creating negative responses to external circumstances that are often beyond their control (Farias & Wikholm, 2016).

Mindfulness in Tourism Research

As in the wider academic and commercial well-being spheres, mindfulness is emerging as a growing concept in tourism research. Overall, there are three categories of mindfulness research in tourism: an emerging set of papers that sit squarely in the McMindfulness or BMM approach, a very small number of studies exploring Buddhist practice among tourists and in tourism planning (cf. Theerapappisit, 2003; Wang *et al.*, 2021), and other studies exploring the applications of LM to a variety of tourism topics.

The more recent papers (cf. Chen *et al.*, 2017; Johnson & Park, 2020; Lengyel, 2020; Stankov *et al.*, 2020) arguing for the various benefits and adoption of the supposedly Buddhist-inspired popular mindfulness trend suffer from all the issues previously listed for McMindfulness. There is confusion over definitions of mindfulness, a variety of false, shallow and incorrect arguments that LM and MBM are different, confusion of meditation with mindfulness, a complete lack of acknowledgement of the issues of cultural appropriation and no critical discussion or even awareness of the available psychology research evidence and its methodological limitations to support their claims about BMM outcomes. It seems that these tourism researchers have simply copied Kabat-Zinn's (1994) definition of mindfulness from another tourism researcher, despite most presenting it as though they have read one of his original popular books, and linked it to Buddhism even though Kabat-Zinn has clearly, and on multiple occasions, stated that his MBM is not Buddhism in anyway. Kabat-Zinn states, '[B]ecause I practice and teach mindfulness, I have the recurring experience that people frequently make the assumption that I am a Buddhist. . . I am not a Buddhist' (Kabat-Zinn, 2014: 343). In a more recent newspaper interview (Booth, 2017: 2) Kabat-Zinn is quoted as saying, 'I bent over backwards to structure it [MBM] and find ways to speak about it that avoided as much as possible the risk of it being seen as Buddhist, new age, eastern mysticism or just plain flakey'. This mistaken representation of Kabat-Zinn's work has led to those few tourism papers that recognise multiple approaches to mindfulness mistakenly claiming that only two mindfulness approaches exist: a sociocognitive and a meditative Buddhism-based approach, which are further incorrectly presented as fundamentally different (cf. Chen *et al.*, 2017). Three groups of researchers in tourism have focused on Kabat-Zinn's 1994 work, ignoring the wider range of psychologists using various forms of meditation combined with cognitive skills training to support mindfulness (Chen *et al.*, 2017; Lengyel, 2020; Stankov *et al.*, 2020a, 2020b). While Kabat-Zinn is something of a celebrity based on popular books he is a molecular biologist interested in medicine, not a psychologist, and the idea of MBM in psychology predates his clinic and celebrity. Arguably discussions of MBM in academic papers should be based on a good understanding of established psychology researchers in MBM including Siegel, Shapiro, Davidson, Creswell, Chiesa, Hanh and Marlatt. Those problematic tourism papers have taken weakly supportive results inappropriately cited in popular, not scholarly, texts and presented them as exciting without any search for actual evidence from properly conducted psychological research and no apparent awareness of the potentially negative aspects of meditation as pathway to mindfulness. It seems that few of these MBM tourism researchers have actually read any of Kabat-Zinn's work or examined analyses of mindfulness in the psychology academic literature; thus the mistaken assumptions and incorrect descriptions keep getting repeated in a downward spiral of increasingly poor scholarship and general confusion.

The more fruitful work in tourism and mindfulness is that conducted within the LM traditions. The author of the present chapter conducted five studies on LM and its implications for learning in museums and protected natural areas for a PhD in sociocognitive psychology (see Moscardo, 1999, for an overview of this work) and subsequently applied it to tourism more generally (see Moscardo, 1996, 1997, 2009, 2017b). These ideas have been taken up by others in research linking mindfulness to recall of positive food tourism experiences (Lee & Kim, 2018) and rural tourism experiences (Loureiro *et al.*, 2019); travel anticipation (Taylor & Norman, 2019) and preference for pro-environmental hotels (Errmann *et al.*, 2021), as well as research confirming the effectiveness of various mindfulness-inducing techniques in cultural heritage presentations (Noor *et al.*, 2015).

Mindfulness in Tourism Practice: Whose Mindfulness and Why Does it Matter?

Three major conclusions can be drawn from the discussion thus far:

(1) The evidence to support meditation as s pathway to mindfulness is at best limited, but there are people who benefit from it and for the time being there is likely to be a market searching for it.
(2) We need to be very careful about the cultural appropriation and commodification of Buddhism as a product for Western tourists.
(3) LM provides the most comprehensive and evidence-based approach to both mindfulness interventions and well-being improvements.

If LM is the overall framework for encouraging greater mindfulness in tourism to enhance well-being, then we need to answer the questions of who should be mindful and why. In terms of the first question, we can identify key stakeholders as tourists, destination residents, tourism businesses and their staff, tourism service providers and tourism planners, policymakers and destination marketing/management organisations. In terms of the second question, mindfulness benefits exist at three levels. The most obvious and immediate is individual well-being, as mindfulness is associated with better cognitive, emotional, creative, learning, decision-making and health outcomes. The second level is the direct consequence of individuals making better decisions and having improved memory and positive emotional responses to situations. Mindful staff can improve business efficiency and performance and mindful tourists can give service providers positive reviews and repeat business. At the third level, and this is connected to the author's values, are the longer term, collective benefits of more effective sustainability messaging to all stakeholders and changes towards actions likely to improve the sustainability of tourism more generally.

Encouraging Mindfulness in Tourism

There are two ways that mindfulness can be used to enhance the well-being of tourism stakeholders. The first is the covert option of encouraging mindfulness in the design of workplace and tourist experience settings using the principles and ideas listed Table 3.1. The second is encouraging mindfulness in any stakeholder group through the overt use of generic mindfulness-based interventions both with and without meditation. The two groups most likely to be targeted with such interventions would be tourism staff and special interest wellness tourists.

For the time being tourists travelling specifically to engage in some form of mindfulness intervention are likely to be seeking an intervention that includes meditation, given the current popularity of McMindfulness. The discussion thus far suggests that tourism businesses selling such programmes need to be careful about how they present their programmes, preferably not commodifying Buddhism as a promotional tool. Such programmes need to be both intensive and extensive and should run for at least 8 to 10 days, with structured and ongoing support for continuing the practices beyond the initial programme and follow up. Programmes should combine cognitive skills training and meditation components should be supervised by qualified counsellors who are able to identify, preferably with formal pre-programme evaluations, and protect people who might be at risk of negative responses to meditation. More details on how to manage such programmes can be found in Siegel (2007, 2020) and various contributions in Ie *et al.* (2014b).

Structured mindfulness programmes for tourism staff, and indeed other groups of stakeholders such as business managers, DMO staff and volunteers, and possibly destination residents, could also be used not only to improve the individual well-being of these stakeholders but also to enhance their decision-making regarding tourism. It would be possible to use mindfulness programmes that include meditation but these need to be treated with caution, as noted in the previous sections. The evidence, however, clearly supports cognitive interventions such as those listed in Table 3.1.

Being Mindful about Mindfulness: Conclusions and Cautions

If we return to the two stories that began this chapter, what is interesting is that it is the dark story, *Total Recall*, that ends in light, a triumph of good over evil. The hero's challenging experiences encourage him to re-evaluate his fundamental beliefs and values and to seek an alternative approach to the problems he faces. In *Eat Pray Love* the film seems to end in light but the story on which the film was based does not. Sadly, the protagonist did not find true love; after returning to her home country the relationship started on a holiday failed and our heroine found another true love with her best female friend. It seems that she didn't find her true self or true love while travelling. These unexpected story endings suggest that we need to be cautious when talking about encouraging mindful tourists as the consequences cannot always be predicted.

We can reasonably argue that encouraging mindfulness should be good for the personal well-being of individual tourists and other stakeholders in tourism. We could also propose that if a tourism provider offers a rewarding service then more mindful visitors should support that business through positive reviews and repeat patronage. We need, however, to be very careful about assuming that a mindful person is an inherently ethical or good person. Serial killers and con artists are arguably very mindful individuals. Mindfulness prepares people to better receive and consider messages related to the things we might value in sustainable tourism, but it is a necessary, not sufficient, condition for action. There are clear links between mindfulness and openness to new information, greater learning, increased ability to take on the perspectives of others and to have empathy for others, changes in attitudes, better alignment between personal values and actions, and increased perceptions of self-efficacy in relation to sustainability actions. All these can create a cognitive state that in tourism is likely to lead to increased support for sustainability action at the individual tourist level and greater demand for sustainability for tourism businesses and management agencies. But this does require that tourists are given information and experiences that not only encourage mindfulness but also offer them guidance on sustainability and opportunities to engage in positive sustainable action while travelling. It also requires that tourism organisations are clear and consistent with their own actions and are trustworthy and reflective on their own practices and values.

We also need to be very aware of the line between cognition and values. Any consideration of mindfulness in tourism requires us as tourism academics and scholars to be mindful of our own values and the gaps between these and our actions. Although the number of tourism research papers that seem to be trapped in the McMindfulness approach is still small, it is increasing. These authors seem to have mindlessly taken the unsubstantiated enthusiasm for BMM exhibited by those who stand to make considerable commercial gains from its uncritical adoption, without any basic scholarly analysis. They present claims and conclusions with little to no supportive evidence and with a failure to seek out, critically read and accurately report on the relevant literature. Mindful tourism cannot be based on this approach. By way of contrast there is a growing body of tourism research exploring LM, although much of this is still focused on identifying how mindfulness is linked to individual tourist decision-making, satisfaction and positive evaluations of their experiences. There is a need to focus research more on evaluating the effectiveness of various mindfulness interventions in changing tourists' sustainability knowledge, concern and action. Tourism scholars are also responsible for educating tourism managers about and for sustainability and they could benefit from the various applications of mindfulness in education.

Returning to my values and beliefs that tourism has to be not just sustainable in its own right but an active contributor to sustainability beyond the tourist experience, I would argue that mindfulness interventions, especially from the LM approach, can encourage greater sustainability. Travel can act as a routine disruption that allows us to be more mindful and opens us to the possibility of challenging our own actions and building better lives at home so that we can travel less but travel better. I would also argue that many elements of mindfulness can be extended to tourism businesses and other organisations (see contributions to the section on organisational mindfulness in Ie *et al.* (2014a) for examples of this type of extension). Mindfulness at a broader level in tourism governance directs us towards empowerment for staff and destination residents, engagement and inclusion of a wider range of stakeholders and continuous and more critical questioning of assumptions and systems that support unnecessary consumption and unsustainable action in tourism.

References

Balcetis, E., Cole, S. and Sherali, S. (2014) The motivated and mindful perceiver. In A. Ie, C. Ngnoumen and E. Langer (eds) *The Wiley Blackwell Handbook of Mindfulness, Volume 1* (pp. 200–215). Chichester: Wiley Blackwell.

Baltzell, A. (ed.) (2016) *Mindfulness and Performance*. New York: Cambridge University Press.

Booth, R. (2017) Master of mindfulness, Jon Kabat-Zinn: 'People are losing their minds. This is what we need to wake up to'. *The Guardian*, 22 October. See www.theguardian.com/lifeandstyle/2017/oct/22/mindfulness-jon-kabat-zinn-depression-trump-grenfell. (accessed May 2021)

Chen, L., Scott, N. and Benckendorff, P. (2017) Mindful tourist experiences: A Buddhist perspective. *Annals of Tourism Research* 64, 1–12.

Chiesa, A. and Serretti, A. (2009) Mindfulness-based stress reduction for stress management in healthy people. *The Journal of Alternative & Complementary Medicine* 15 (5), 593–600.

Cook, J. (2016) Mindful in Westminster. *HAU: Journal of Ethnographic Theory* 6 (1), 141–161.

Errmann, A., Kim, J., Lee, D.C., Seo, Y., Lee, J. and Kim, S.S. (2021) Mindfulness and pro-environmental hotel preference. *Annals of Tourism Research* 90. https://doi.org/10.1016/j.annals.2021.103263.

Evans, J.S.B.T. (2018) Dual-process theories. In L.J. Ball and V.A. Thompson (eds) *The Routledge International Handbook of Thinking and Reasoning* (pp. 151–166). London: Routledge.

Farias, M. and Wikholm, C. (2016) Has the science of mindfulness lost its mind? *BJPsych Bulletin* 40 (6), 329–332.

Fatemi, S. (ed.) (2016a) *Critical Mindfulness*. Cham: Springer.

Fatemi, S. (2016b) Critical mindfulness of psychology's mindlessness. In S. Fatemi (ed.) *Critical Mindfulness* (pp. 1–24). Cham: Springer.

Fischer, D., Stanszus, L., Geiger, S., Grossman, P. and Schrader, U. (2017) Mindfulness and sustainable consumption: A systematic literature review of research approaches and findings. *Journal of Cleaner Production* 162, 544–558.

Geiger, S.M., Grossman, P. and Schrader, U. (2019) Mindfulness and sustainability. *Current Opinion in Psychology* 28, 23–27.

Gethin, R. (2011) On some definitions of mindfulness. *Contemporary Buddhism* 12 (1), 263–279.

Helm, S. and Subramaniam, B. (2019) Exploring socio-cognitive mindfulness in the context of sustainable consumption. *Sustainability* 11 (13). https://doi.org/10.3390/su11133692.

Herbert, W. (2014) Mindfulness and heuristics. In A. Ie, C. Ngnoumen and E. Langer (eds) *The Wiley Blackwell Handbook of Mindfulness, Volume 1* (pp. 279–289). Chichester: Wiley Blackwell.

Hickey, W.S. (2010) Meditation as medicine: A critique. *CrossCurrents* 60 (2), 168–184.

Hölzel, B.K., Lazar, S.W., Gard, T., Schuman-Olivier, Z., Vago, D.R. and Ott, U. (2011) How does mindfulness meditation work? *Perspectives on Psychological Science* 6 (6), 537–559.

Hyland, T. (2017) McDonaldizing spirituality: Mindfulness, education, and consumerism. *Journal of Transformative Education* 15 (4), 334–356.

Ie, A., Ngnoumen, C. and Langer, E. (eds) (2014a) *The Wiley Blackwell Handbook of Mindfulness, Volume 1*. Chichester: Wiley Blackwell.

Ie, A., Ngnoumen, C. and Langer, E. (eds) (2014b) *The Wiley Blackwell Handbook of Mindfulness, Volume 2*. Chichester: Wiley Blackwell.

Ivtzan, I. and Hart, R. (2016) Mindfulness scholarship and interventions. In A. Baltzell (ed.) *Mindfulness and Performance* (pp. 3–28). Boston: Cambridge University Press.

Jamieson, S.D. and Tuckey, M.R. (2017) Mindfulness interventions in the workplace. *Journal of Occupational Health Psychology* 22 (2), 180–193.

Johnson, K.R. and Park, S. (2020) Mindfulness training for tourism and hospitality frontline employees. *Industrial and Commercial Training* 52 (3), 185–193.

Kabat-Zinn, J. (1994) *Wherever You Go, There You Are: Mindfulness Meditation in Everyday Life*. New York: Hachette/Hyperion.

Kabat-Zinn, J. (2005) *Coming to Our Senses: Healing Ourselves and the World through Mindfulness*. New York: Hachette/Hyperion.

Kabat-Zinn, J. (2014) Meditation is not for the faint-hearted. *Mindfulness* 5 (3), 341–344.

Kang, Y., Gruber, J. and Gray, J. (2014) Mindfulness. In A. Ie, C. Ngnoumen and E. Langer (eds) *The Wiley Blackwell Handbook of Mindfulness, Volume 1* (pp. 168–185). Chichester: Wiley Blackwell.

Kanter, R. and Fox, D. (2016) Understanding confidence. In S. Fatemi (ed.) *Critical Mindfulness* (pp. 55–68). Cham: Springer.

Langer, E. J. (2000) Mindful learning. *Current Directions in Psychological Science* 9 (6), 220–223.

Langer, E. (2014) Mindfulness forward and backward. In A. Ie, C. Ngnoumen and E. Langer (eds) *The Wiley Blackwell Handbook of Mindfulness, Volume 1* (pp. 7–21). Chichester: Wiley Blackwell.

Lee, S. (2019) Psychology's own mindfulness. *Journal of the History of the Behavioral Sciences* 55 (3), 216–229.

Lee, Y. and Kim, I. (2018) Investigating key innovation capabilities fostering visitors' mindfulness and its consequences in the food exposition environment. *Journal of Travel & Tourism Marketing* 35 (6), 803–818.

Lengyel, A. (2015) Mindfulness and sustainability: Utilizing the tourism context. *Journal of Sustainable Development* 8 (9), 35–51.

Lengyel, A. (2020) Authenticity, mindfulness and destination liminoidity. *Tourism Recreation Research* 47 (1), 31–46. https://doi.org/10.1080/02508281.2020.1815412.

Loureiro, S.M.C., Breazeale, M. and Radic, A. (2019) Happiness with rural experience. *Journal of Vacation Marketing* 25 (3), 279–300.

Marks, D.R. and Cowan, E. (2017) Langerian explorations. *PsycCRITIQUES* 62 (4). DOI: 10.1037/a0040692.

Moldoveanu, M. (2016) The construct of mindfulness amidst and along conceptions of rationality. In S. Fatemi (ed.) *Critical Mindfulness* (pp. 25–44). Cham: Springer.

Moscardo, G. (1996) Mindful visitors: Heritage and tourism. *Annals of Tourism Research* 23 (2), 376–397.

Moscardo, G. (1997) Making mindful managers. *Journal of Tourism Studies* 8 (1), 16–24.

Moscardo, G. (1999) *Making Visitors Mindful*. Champaign: Sagamore Publishing.

Moscardo, G. (2009) Understanding tourist experience through mindfulness theory. In M. Kozak and A. Decrop (eds) *Handbook of Tourist Behavior* (pp. 99–115). New York: Routledge.

Moscardo, G. (2017a) Critical reflections on the role of interpretation in visitor management. In J. Albrecht (ed.) *Visitor Management in Tourism Destinations* (pp. 170–187). Wallingford: CABI.

Moscardo, G. (2017b) Exploring mindfulness and stories in tourist experiences. *International Journal of Culture, Tourism & Hospitality Research* 11 (2), 111–124.

Niculescu, M. (2020) 'Jewish mindfulness' as spiritual didactics teaching Orthodox Jewish religion through mindfulness meditation. *Religions* 11 (1), 11.

Niedderer, K. (2014) Mediating mindful social interactions through design. In A. Ie, C. Ngnoumen and E. Langer (eds) *The Wiley Blackwell Handbook of Mindfulness, Volume 1* (pp. 345–366). Chichester: Wiley Blackwell.

Nilsson, H. and Kazemi, A. (2016) Reconciling and thematizing definitions of mindfulness. *Review of General Psychology* 20 (2), 183–193.

Noor, S., Rasoolimanesh, S.M., Ganesan, V. and Jaafar, M. (2015) Effective interpretation using various media toward mindfulness. *Journal of Heritage Tourism* 10 (3), 263–279.

Oman, D. (2020) Studying the effects of meditation. In M. Farias, D. Brazier and M. Lalljee (eds) *The Oxford Handbook of Meditation* (pp. 41–75). Oxford: Oxford Academic. DOI: 10.1093/oxfordhb/9780198808640.001.0001.

Pagnini, F. and Phillips, D. (2015) Being mindful about mindfulness. *The Lancet Psychiatry* 2 (4), 288–289.

Poceski, M. (2020) Mindfulness, cultural appropriation, and the global diffusion of Buddhist contemplative practices. *International Journal for the Study of Chan Buddhism and Human Civilization* 7. See http://cbhc.crs.cuhk.edu.hk/main/wp-content/uploads/2020/03/E-journal_-Issue-7_Mario-Poceski.pdf. (accessed May 2021)

Purser, R.E. (2018) Critical perspectives on corporate mindfulness. *Journal of Management, Spirituality & Religion* 15 (2), 105–108.

Saeed, S.A., Antonacci, D.J. and Bloch, R.M. (2010) Exercise, yoga, and meditation for depressive and anxiety disorders. *American Family Physician* 81 (8), 981–986.

Scherer, B. and Waistell, J. (2017) Incorporating mindfulness: Questioning capitalism. *Journal of Management, Spirituality & Religion* 15 (5), 1–18.

Siegel, D. (2007) *The Mindful Brain*. New York: W.W. Norton.

Siegel, D.J. (2020) *The Developing Mind* (3rd edn). New York: Guilford Press.

Siegling, A.B. and Petrides, K.V. (2014) Measures of trait mindfulness. *Frontiers in Psychology* 5. https://doi.org/10.3389/fpsyg.2014.01164.

Stankov, U., Filimonau, V. and Vujičić, M.D. (2020a) A mindful shift. *Tourism Geographies* 22 (3), 703–712.

Stankov, U., Filimonau, V., Gretzel, U. and Vujičić, M.D. (2020b) E-mindfulness – the growing importance of facilitating tourists' connections to the present moment. *Journal of Tourism Futures* 6(3), 239–245.

Tan, C. (2021) Confucius and Langerian mindfulness. *Educational Philosophy and Theory* 53 (9), 831–940.

Taylor, L.L. and Norman, W.C. (2019) The influence of mindfulness during the travel anticipation phase. *Tourism Recreation Research* 44 (1), 76–90.

Theerapappisit, P. (2003) Mekong tourism development: Capital or social mobilization? *Tourism Recreation Research* 28 (1), 47–56.

Thiermann, U.B. and Sheate, W.R. (2021) The way forward in mindfulness and sustainability. *Journal of Cognitive Enhancement* 5 (1), 118–139.

Thomas, J., Furber, S.W. and Grey, I. (2017) The rise of mindfulness and its resonance with the Islamic tradition. *Mental Health, Religion & Culture* 20 (10), 973–985.

Thomas, S. (2015) *The Seed Collectors*. New York: Soft Skull Press.

Tisdell, E.J. and Riley, T.D. (2019) The landscape of mindfulness and meditation in adult education. *New Directions for Adult and Continuing Education* 161, 9–20.

Trammel, R.C. (2017) Tracing the roots of mindfulness. *Journal of Religion & Spirituality in Social Work: Social Thought* 36 (3), 367–383.

Van Dam, N.T., Van Vugt, M.K., Vago, D.R., Schmalzl, L., Saron, C.D., Olendzki, A., Meissner, T., Lazar, S.W., Kerr, C.E., Gorchov, J., Fox, K.C.R., Field, B.A., Britton, W.B., Brefczynski-Lewis, J.A. and Meyer, D.E. (2018) Mind the hype: A critical evaluation and prescriptive agenda for research on mindfulness and meditation. *Perspectives on Psychological Science* 13 (1), 36–61.

Wamsler, C. and Brink, E. (2018) Mindsets for sustainability. *Ecological Economics* 151, 55–61.

Wang, Y.C., Chen, P.J., Shi, H. and Shi, W. (2021) Travel for mindfulness through Zen retreat experience. *Tourism Management* 83. https://doi.org/10.1016/j.tourman.2020.104211.

4 Volunteer Tourism: A Pathway to Hope and Happiness?

Reni Polus and Neil Carr

Introduction

Like any kind of tourism, volunteer tourism can have a diverse range of impacts, particularly on the community where tourism takes place. Many scholars have praised volunteer tourism for its contribution to the welfare of communities (Coghlan & Weiler, 2018; Pan, 2012; Wearing, 2001; Zahra, 2010; Zahra & McGehee, 2013). Nevertheless, some claim that this form of tourism has adopted the mass tourism model of packaging and segmentation, with consequent negative implications for host communities (Tomazos & Butler, 2009). This scenario was noted by Guttentag (2009), who pointed out a variety of potential adverse effects that can be caused by international volunteer tourism. Perhaps the most frequent criticism of volunteer tourism has been whether it truly embodies the concept of altruism, which is meant to be central to Western conceptualisations of volunteering (Cnaan et al., 1996). Where the greatest benefits of volunteer tourism experiences accrue to the volunteer rather than the host communities, the altruistic motive of international volunteer tourists has been questioned (Coghlan, 2015; Coghlan & Fennell, 2009; Fennell, 2006; Guttentag, 2009; Holmes et al., 2007). Viewing volunteer tourists as only motivated by altruism is a very superficial position to adopt as social and hedonistic benefits will always be something that volunteer tourists seek or gain from their experiences. Recognition of this means that consensus on the nature of altruism in volunteer tourism has yet to be achieved (Coghlan & Fennell, 2009; Wright, 2013). Indeed, Coghlan and Fennell (2009: 393) state that 'the role that altruism might play in volunteer tourism is unclear'. This lack of clarity potentially affects how the industry is perceived.

Another limitation of current understandings and representations of volunteer tourism is that they have focused almost exclusively on economic and social aspects of the industry, ignoring subjective aspects in the process. Indeed, despite the prominence of the notion of well-being in tourism studies (e.g. Smith & Diekmann, 2017; Pope, 2018), this concept is rarely discussed by academics or industry operators in the context of volunteer tourism. This is problematic as understanding the effects of volunteer tourism experiences on volunteer tourists' and communities' personal well-being, beyond economic and social issues, will provide critical insights into the impacts of volunteer tourism. This chapter provides a conceptual contribution to the idea of reciprocal altruism in volunteer tourism and considers the predictions this theory makes about personal well-being among both volunteers and the communities they visit. That is, rather than limiting the enquiry by seeking to understand reciprocal altruism in volunteer tourism experiences as ends in themselves, this chapter considers how reciprocal altruism might engender hope that may promise happiness and well-being among both volunteers and the communities they visit.

The chapter begins by providing an overview of definitions and theorisations about altruism, highlighting the sociobiological theory of reciprocal altruism. The discussion then turns to examining the literature on altruism in volunteer tourism experiences. It should be noted that the goal of this chapter is not to provide an inclusive overview of all the various positions on the phenomenon of altruism, or to provide a comprehensive analysis of all extant work on altruism in volunteer tourism studies. The reader is directed to fine examples of such work that already exist if they wish to view such an overview (see Batson, 1991; Bykov, 2016; Clavien & Chapuisat, 2013; Coghlan & Fennell, 2009; Fennell, 2006; Hoffman, 1978; Piliavin & Charng, 1990; Sin, 2010). Instead, this chapter provides an overview of reciprocal altruism in order to build on it to examine the potential for hope and happiness through volunteer tourism. Consequently, having presented a discussion of altruism in volunteer tourism, the chapter then discusses the possibilities of and for reciprocal altruism in volunteer tourism creating hope and happiness, which are conceptualised as both pursuits and outcomes.

The Notion of Altruism

The term 'altruism' was introduced by Auguste Comte, a French philosopher who is also known as the founder of sociology and positivism. He defined altruism as the desire or tendency to 'live for others' (Comte, 1853, cited in Del Pozo, 2002: 40). Comte (1853) coined the term altruism in juxtaposition to egoism. It subsequently became a prominent concept within a variety of academic disciplines and fields of study, including psychology, sociology, philosophy, economics, sociobiology and tourism. Due to the wide use of the concept and the demands of scholars for a precise meaning, defining what is meant by altruism is not an easy task. Nevertheless, there have been frequent attempts to construct a single definition of altruism, resulting in a range of different approaches and potentially adding to rather than resolving this definitional problem. Clavien and Chapuisat (2013: 134) note, '[T]he notion of altruism has become so plastic that it is often hard to understand what is really meant by the authors using the term'. The plasticity of altruism has been demonstrated in several studies, including tourism-focused ones, showing how it is affected by the context in which it is set and viewed through.

When viewed from a psychology perspective, altruism can be identified as a product of socialisation. As Haski-Leventhal (2009: 276) argues, '[P]eople are born egoistic and change through socialization and [upbringing].' Accordingly, altruism is intimately linked with the ideas of learning and growing as humans are naturally egoistic (Haski-Leventhal, 2009; Piliavin & Charng, 1990). These claims have been strongly contested by several scholars. Critics have argued that humans are programmed as both egoistic and altruistic (Hoffman, 1978; Krebs & Van Hesteren, 1994; Zahn-Waxler *et al.*, 1992). As children grow, they develop altruistic behaviours not only due to environmental factors, such as socialisation or parental modelling, but also due to temperament factors, such as personal character or personality development (Zahn-Waxler *et al.*, 1992). The notion of nurture versus nature is integral to this debate where altruism is conceptualised as either biological or cultural transmittance, or potentially a blend of the two.

Batson (1991: 6) defines altruism as 'a motivational state with the ultimate goal of increasing another's welfare'. This definition stresses the interests of others as the foundational goal of the altruistic action, which differentiates the conceptualisation of altruism from egoism or self-preserving behaviours. Another prominent definition of altruism was provided by Monroe (1998: 6), who defined it as 'behaviour intended to benefit another, even when this risks possible sacrifice to the welfare of the actor'. Monroe examined the possible influences that encourage altruism.

To make things more complicated, sociologists have conceptualised altruism as a form of prosocial behaviour, alongside helping and cooperation (Aydinli *et al.*, 2013). However, prosocial behaviour more accurately refers to the consequences of a person's actions while altruism is the motivation behind the action (Afolabi, 2013). In an economic paradigm, altruism has been viewed as a 'product with a price tag . . . with a potential profit' (Haski-Leventhal, 2009: 284). From this perspective, altruism has a calculable economic value in the form of human capital investment (Knox, 1999; Monroe, 1998). Economists also suggest that self-rewards are the ultimate goal of acting altruistically (Knox, 1999). Both concepts stand in opposition to the notion of altruism set out in other disciplines; the term altruism is described as the consequences of an action in hope of future benefit.

Returning to a psychological perspective, one question that needs to be asked is whether altruism exists naturally in humans. A sociological perspective would say yes, but at least some psychologists clearly disagree.

The aforementioned theories provide useful explanations of the concept of altruism but do not explain why humans act altruistically (Fennell, 2006; Hamilton, 1964; Trivers, 1971). Over the last few decades, sociologists have used the concepts of proximate and ultimate causes of altruism in humans. Proximate causation refers to altruistic behaviour that is influenced by personality dispositions, empathy, guilt, and social norms. In comparison, altruistic behaviour that is driven by individuals selectively cooperating with members of their in-group is referred to as ultimate causation (Batson, 1991; Hamilton, 1964; Trivers, 1971).

As such, altruism has also been studied through a sociobiological lens to explain altruistic behaviour as one method of insuring gene propagation without having to deal with moral questions. In this context, Hamilton (1964) provided theoretical guidance through his theory of inclusive fitness. He recognised that altruism could be explained through kinship genetics where humans pass their genetic information on to successive generations. In other words, humans extend altruism to others as a function of relatedness. Love, care, sympathy and compassion are some of the underlying intentions of this form of altruistic behaviour (Curry *et al.*, 2018). However, the main weakness of this theory is that it does not appear to account for altruism between non-related individuals.

To explain how and why altruistic human social behaviour would take place between unrelated individuals, Trivers (1971) developed the concept of reciprocal altruism. He presented altruism as a form of symbiosis where actors help one another with the expectation of a return favour at a later date (Trivers, 1971). Thus, there is reciprocity, or a cooperative relationship, between individuals, though immediate reciprocity is not necessarily required. The basic idea of cooperation is that it requires a positive outcome for both parties (Aydinli *et al.*, 2013), which is a requisite outcome of reciprocity (Dovidio *et al.*, 2017). Sympathy, trust, returning favours, gratitude, forgiveness and friendship are noted as the underlying intentions of reciprocal altruism (Curry *et al.*, 2018).

Trivers's (1971) reciprocal altruism theory has not escaped criticism. The expectation of the returned favour is viewed by many as the underlying goal. This is tied to one's individual well-being. This makes it a highly egoistical behaviour rather than an altruistic one (Batson, 1991; Cialdini *et al.*, 1987; Hoffman, 1981; Mayr, 1988). In response to this critique, Fennell (2006) argued that the essence of reciprocal altruism is human behaviour. As he put it, 'although we are very much a self-interested species, we are also a moral one' (2006: 121), suggesting an altruistic heart beats at the core of reciprocal behaviour that may include a component of egoism. This view is consistent with that of Trivers (1971), who claimed that the ability to distinguish altruists from egoists and the willingness of the individual to reciprocate is a prerequisite for the evolution of altruism. Reciprocal altruism is deemed relevant to mutual support and social harmony (Lai *et al.*, 2012; Wentzel, 1998) because individuals who carry out altruistic acts understand they will benefit from their good deeds as well. Reciprocal altruism explains how conditionally helping others can potentially benefit the long-term interests of each helper, giving them hope and potentially happiness, which is, of course, linked to well-being (Becchetti *et al.*, 2016; Staats & Stassen, 1985).

In relation to tourism studies, Fennell (2006) provided a conceptual piece on reciprocal altruism to understand the motivations of altruistic behaviour between hosts and tourists. The key argument is that '[collective] interaction between many hosts and guests provides the possibility for an emerging symbiotic relationship over time' (2006: 118). Instead of self-interest, Fennell's reciprocal altruism suggests that the goodwill between hosts and guests *might* – hence the position of hope – create a climate of mutually beneficial cooperation. To date, only a few studies have empirically examined this phenomenon (Jacob & Guéguen, 2012; Uriely *et al.*, 2002; Woosnam & Norman, 2010). Before looking further into the possibilities of reciprocal altruism in volunteer tourism, the next section will discuss where altruism has, up until now, been referred to within volunteer tourism studies.

Altruism and Volunteer Tourism

Over the years, volunteer tourism has been linked with a variety of contexts and perspectives. In particular, it has been identified as a form of 'alternative tourism'. This definition and the trend of people towards it have been identified as a response to the negative impacts of mass tourism (Björk, 2000; Cohen, 1987; Singh, 2002; Uriely *et al.*, 2003; Wearing, 2001), making it an inherently hopeful experience, in that it hopes to do better than mass tourism for all involved. Volunteer tourism has also been associated with ecotourism, responsible tourism and sustainable tourism (Brown, 2005; Callanan & Thomas, 2005; Wearing, 2001), all of which similarly exist and are defined in counterpoint to the negative images associated with mass tourism. The argument is that volunteer tourism appears to be able to offer an alternative direction, where profit-driven objectives are secondary to a more altruistic desire to travel in order to assist the community (Wearing, 2001). In this context, Wearing (2001) regards the volunteer tourist as someone who pays to travel to another location where they choose to use their free time to engage in a meaningful experience by volunteering. He defined this concept as a trip that involves 'aiding or alleviating the material poverty of some groups in society, the restoration of certain environments or research into aspects of society or environment' (2001: 1). For Singh and Singh (2001, cited in Singh, 2004: 174), volunteer tourism is a form of altruistic tourism: '[It is] more of a conscientious practice of righteous tourism – one that comes closest to utopia. At best, it may be regarded as an altruistic form of tourism, which has the capacity to uphold the highest ideals, intrinsically interwoven in the tourism phenomenon'.

Unsurprisingly, altruism is commonly associated with volunteer tourism as a motivational force for participants. It has often been used to distinguish volunteer tourists from other types of tourists. Indeed, Mustonen (2006: 165) noted that 'volunteer tourists differ from other tourists when the motivation basis is concerned. For them the main motivation is linked with an altruistic desire to volunteer'. However, several scholars have pointed out the need to rethink the concept of altruism in volunteer tourism, as self-interest has also been noted as part of the motivational force (Coghlan & Fennell, 2009; Knollenberg et al., 2014; Wearing & McGehee, 2013). In this context, the altruistic motive of international volunteer tourists has been questioned, with the idea that there is no pure, selfless altruism in volunteer tourism. Rather, it is argued that volunteer tourists who are altruistically helping others are also benefiting from various rewards experienced during the process. Indeed, the altruism versus egoism debate that was highlighted in the previous section has been extensively researched within volunteer tourism, resulting in a range of frameworks (Brown, 2005; Coghlan & Fennell, 2009; Daldeniz & Hampton, 2010; Lee & Yen, 2015; Mustonen, 2006).

Nevertheless, it is argued that when the actual altruistic behaviour of the volunteer tourist is considered, in addition to altruism and egoism dimensions, there is another dimension that has been overlooked: reciprocity (Callanan & Thomas, 2005; Coghlan & Fennell, 2009). Recognising this, it is important to explore the chances for volunteer tourism encounters to achieve cooperation. Thus, there are reciprocal altruism theoretical approaches that are important to consider in disentangling the issues around volunteer tourism's motivations and ethics. Scholars suggest that volunteer tourism represents, or at least has the potential to represent, a form of reciprocal altruism (Ooi & Laing, 2010; Paraskevaidis & Andriotis, 2017; Söderman & Snead, 2008). For example, Söderman and Snead (2008: 123) note, '[I]t is clear that volunteering is not just about "doing good" for others but it is also about doing good for self, about reciprocal altruism'. Similarly, McIntosh and Zahra (2008b: 166), who researched a volunteer tourism project situated in a New Zealand Māori community, introduced their study by stating, '[I]mportantly for this chapter, volunteer tourism is seen to foster a reciprocal and mutually beneficial relationship between the host and guest'. However, empirical research into the reciprocal altruism of volunteer tourism experiences remains noticeably lacking. It is with this point in mind that we need to ask, what are the possibilities of and for reciprocal altruism in volunteer tourism, and what are the chances of this collective interaction fostering positive personal well-being and happiness for everyone involved? It is the *chances* of this happening that form the foundation of hope in relation to happiness and well-being for hosts and guests through volunteer tourism. What follows is an attempt to more fully ground the ideas behind reciprocal altruism in volunteer tourism.

The Possibilities of and for Volunteer Tourism as a Pathway to Hope and Happiness

Altruism is widely recognised as the basis for volunteer tourism. However, this notion has received numerous critiques as there is no pure, selfless altruism in volunteer tourism (Coghlan & Fennell, 2009). Moving beyond this mono-altruism, this chapter identifies the presence and potential of reciprocal altruism in volunteer tourism, conceptualised as both pursuits and outcomes for both hosts and guests. Significantly, this chapter suggests that volunteer tourism has the potential to create hope that may promise happiness and well-being among both volunteers and the communities they visit. However, there is no one-size-fits-all response to volunteer tourism, recognising the reality that it may be imposed on the local community, potentially resulting in negative perceptions or even resentment towards the volunteer tourists. Hope and happiness are closely connected but differ in terms of temporal orientation: hope is directed towards the future while happiness is oriented towards the here and now (Bailey et al., 2007; Genda, 2016; Witvliet et al., 2019). Hope arguably has the potential to shift beyond the simplicity of a cost–benefit analysis within reciprocal altruism in novel and unique ways. It can offer a nuanced, qualitative way of approaching reciprocal altruism that is lacking a cost–benefit analysis.

In taking a goal pursuit perspective, collective altruism and sustained selflessness are the foundation of the altruistic behaviour of volunteer tourists. Collective altruism is referred to as 'a reliance on the goodwill of others to accomplish collective, altruistic ends' (Fennell, 2006: 118). A likely site to find such collective forms of altruism in a tourism setting is volunteer tourism, where there is a collective commitment to acting altruistically. However, recognising that volunteer tourism and volunteer tourists are different in many ways, with the former a concept and the latter consisting of unique individuals, not all volunteer tourists may fit this mould. Within this context, collective altruism is expressed through sustained selflessness. Fennell (2006: 118) notes that '[volunteer] tourists will not only be helping other tourists who come into contact with a given host, but also helping themselves through the expectation that another tourist has acted altruistically before for his or her benefit'. Following this view, it can be argued that volunteer tourism is a form of tourism where individuals act altruistically to contribute to collective benefits. Several lines of evidence have demonstrated the reciprocal and mutually beneficial relationship between volunteer tourists and members of host communities (Callanan & Thomas, 2005; McIntosh & Zahra, 2008a; Sin, 2009, 2010; Tomazos & Cooper, 2012). In this vein, volunteer tourists develop a sense of hope that their action is a contribution that may continue to live through others. Indeed, McGeer (2004: 104) argues that hope 'is a matter, not only of recognizing but also of actively engaging with our own current limitations in affecting the future we want to inhabit'. In other words, hope is a crucial and strong motivational force that is seen as 'the fuel for agency' (Lueck, 2007: 256) to act even in the absence of certainties (Courville & Piper, 2004; McGeer, 2004). This aspect of the experience is perhaps most significant not only for the volunteer tourists who desire change and transformation but also for the local communities who view volunteer tourism as a source of potential change.

From an outcome perspective, collective altruism in volunteer tourism increases the personal well-being of hosts and tourists. Within this, happiness is both an indicator and a driver of well-being. Evidence shows that altruistic behaviour and happiness are interconnected (Brown *et al.*, 2003; Taylor & Turner, 2001). Not only can performing altruistic actions cause happiness in itself, but the results of such actions can also lead to happiness for oneself and others. However, with the ideological essence of volunteer tourism focusing on contributions to host communities in the form of environmental, economic and sociocultural benefits, little is known about any potential benefits in terms of personal well-being and happiness to both hosts and volunteer tourists.

As noted earlier, volunteer tourists are often portrayed as self-interested. Yet even where volunteer tourists fit this notion, they may still be influenced by the close local connections and interactions with other volunteers that they experience. This may lead to a change in values, which can influence the personal well-being and happiness of both sides. This recognises that values and motivations alter over the life course rather than being fixed and that the tourism experience, occurring in a hedonic space, offers potentially fertile ground for change even across relatively small periods of time. As such, the volunteer tourism space and time offer a potentially fertile ground for producing meaningful relationships from individual and group endeavours and interactions, which may also contribute to personal well-being. Research has shown that there is a link between reciprocal altruism and responsible tourism (Fennell, 2008).

It is becoming apparent that altruism in volunteer tourism is a form of reciprocity where both the host community and volunteers influence each other and the encounter they produce. Furthermore, it is clear that, while issues of self-interest may occupy both volunteer tourists and hosts, this does not negate the position and importance of reciprocal altruism in and through the experience. The frequency and quality of volunteer tourism host–tourist encounters allow collective benefits to unfold. Within mainstream mass tourism, the host–tourist encounter, in general, is brief, superficial and non-repetitive (Reisinger & Dimanche, 2010; Reisinger & Turner, 1998). As a result, tourists often have limited opportunity, or desire, to develop significant relationships with hosts, and potentially vice versa (Reisinger & Dimanche, 2010). Thus, this relationship is often viewed as a form of exploitation, laced with deceit, mistrust and unequal power dynamics (Krippendorf, 1987; Reisinger & Dimanche, 2010). Attempting to overcome this, Fennell (2006: 118) suggests that:

Successive reciprocal altruism [an emerging theory within studies] depicts consecutive generations of tourists moving in and out of touristic space over short periods of time, along with their interaction with specific members of the host region. The repeated nature of primarily altruistic interactions between tourists and host may provide a foundation for achieving sustained levels of trust and cooperation that withstand the test of time.

Studies have demonstrated that the length of stay and travel motivation of volunteer tourists can influence the intensity of host–tourist interactions (Gray & Campbell, 2007; Kirillova *et al*., 2015). In contrast to mass tourists, volunteer tourists normally stay longer in one place, a situation that provides the opportunity for profound social contact with the local community (Kirillova *et al*., 2015). Also, studies suggest that because volunteer tourists often work side-by-side with members of the local community, the encounter has the potential to be authentic, mutually beneficial, and more sustainable than a conventional leisure trip in a mass tourism setting (Gray & Campbell, 2007). This may be reinforced through the opportunities volunteer tourists have to go into local people's homes and see and do things that many other tourists do not get, or probably even want, to do. As such, this chapter argues that in volunteer tourism settings the collective host–tourist interaction may provide a 'foundation for achieving sustained levels of trust and cooperation' (Fennell, 2006: 118), providing, in the process, fertile ground for the development of hope and happiness. The evidence presented here suggests that the intense host–tourist encounters between volunteer tourists and the local community create a climate of cooperation that may result in a symbiotic relationship that benefits both sides, particularly in terms of their personal well-being and happiness.

Indeed, a number of studies have postulated a convergence between volunteer tourism and the personal well-being of both hosts and tourists (Lee, 2020a, 2020b; Zahra & McIntosh, 2007). One notable study was done by Lee (2020a). Looking from a positive psychology perspective, Lee (2020a) examines both hedonic and eudaimonic happiness in the volunteer tourism context, focusing on the impact on individuals in host communities. The findings show that volunteer tourism does not simply bring instant happiness and pleasure to local communities, but, with the assistance and aspirations of volunteers, local communities' eudaimonic happiness also improves. The host–tourist interactions help to facilitate the personal growth of individuals in the long run. As such, the ideal outcomes of reciprocal altruism in the volunteer tourism setting are based on making a positive contribution to not only the social, natural and economic environment in which the experience is situated but also the subjective aspects of life, where both the volunteer and members of the host community gain from the experience. Here we see the potential for the mutually beneficial, rather than conflicting, blending of altruism and egoism. Accordingly, reconceptualising reciprocal altruism in volunteer tourism, this chapter proposes that there is a need to shift the focus away from the tangible outcomes of volunteer tourism (i.e. what has been changed in/for the local community or the volunteer tourists). Rather, we should look at the qualities of outcomes of the collective interaction of volunteer tourism on the personal well-being and happiness of both sides. Volunteer tourism carries potential hope and happiness for both volunteer tourists and hosts, thereby potentially creating reciprocal happiness.

Conclusion

This chapter offers insight into the possibilities of and for reciprocal altruism in volunteer tourism to generate reciprocal happiness and well-being in local communities and volunteer tourists. Consensus on the nature of altruism in volunteer tourism has yet to be achieved, which may just reflect a complex reality that overly simplistic models are never going to be able to accurately depict. This is to suggest that a reductionist quasi-positivistic mindset is a poor lens through which to look when seeking to understand and appreciate the complex nuances of human motivations and behaviour. Consequently, this chapter recognises that altruism in volunteer tourism is a complex idea that needs to be explained by using a multidisciplinary approach, as proposed by Fennell (2006), that embraces this complexity rather than trying to simplify it.

This chapter identifies the presence and potential of reciprocal altruism in volunteer tourism, conceptualised as both pursuits and outcomes. In this context, the altruistic behaviour of volunteer tourists is a form of sustained selflessness that aims to create a climate of collective altruism. In taking a goal pursuit perspective, when altruistic behaviours are considered, it is rather superficial to look just from altruism and egoism perspectives, as altruistic behaviour results in a return of both tangible and intangible benefits. Based on superficial approaches the reasons for volunteer tourists acting altruistically have typically been identified as 'to help others' (altruism) and 'to help oneself' (self-interestedness). This chapter suggests there is a need to move beyond this to recognise that egoism and altruism are interwoven in potentially complementary ways rather than being distinctly separate and conflicting.

From an outcome perspective, altruism in volunteer tourism is a form of reciprocity where both the host community and volunteers influence each other and the encounters they produce. This form of host–tourist interaction is linked to the concept of the 'mutual gaze', where both the volunteer tourist and local gazes coexist (Maoz, 2006). In other words, the local community not only helps to construct the experience of the volunteer tourist but also directly benefits from it, and vice versa. In this case, reciprocal altruism can be observed in host–tourist cooperative traits that are aimed at achieving mutual instant and/or potential future benefits, which is where hope, and the related phenomenon of trust, is located. This recognises that the mutuality lies in the idea of benefits rather than the specificities of them, with hosts and guests potentially gaining different benefits but all deriving a level of happiness, understanding there is no *one* happiness. Recognising this, it is clear that the essential component of reciprocal altruism in volunteer tourism is the host–tourist interaction.

This chapter recognises that understanding the significance of personal well-being and happiness in volunteer tourism within a reciprocal altruism framework that recognises the interlinking nature of egoism and altruism will enable a more holistic and balanced conceptualisation of this segment of the tourism experience to be developed. It can help to critically examine how volunteer tourism can be an effective tool for furthering inclusive and sustainable development, while also aiding the personal well-being of volunteer tourists. It may also help to provide a nuanced and more realistic understanding of what a volunteer tourist really is and can be. Overall, this chapter provides a theoretical foundation that can be a starting point for empirical studies of reciprocal altruism, well-being, and happiness in volunteer tourism. Based on the ideas explored in this chapter, future research could explore the role of host–guest interaction in cocreating tourism experiences and the role that hope, happiness, and reciprocal altruism plays, and has the potential to play, in this. Such work can be situated in multiple volunteering contexts, both inside and beyond tourism. In addition, it will be interesting to assess the potential for the ideas forwarded in this chapter to function and be of value outside of the volunteerism setting.

References

Afolabi, O.A. (2013) Roles of personality types, emotional intelligence and gender differences on prosocial behavior. *Psychological Thought* 6, 124–139. https://doi.org/10.23668/psycharchives.1918.

Aydinli, A., Bender, M. and Chasiotis, A. (2013) Helping and volunteering across cultures: Determinants of prosocial behavior. *Online Readings in Psychology and Culture* 5, 1–27.

Bailey, T.C., Eng, W., Frisch, M.B. and Snyder, C. (2007) Hope and optimism as related to life satisfaction. *The Journal of Positive Psychology* 2, 168–175.

Batson, C.D. (1991) *The Altruism Question: Toward a Social-Psychological Answer.* Washington DC: Lawrence Erlbaum Associates, Inc.

Becchetti, L., Corrado, L. and Conzo, P. (2016) Sociability, altruism and well-being. *Cambridge Journal of Economics* 41, 441–486.

Björk, P. (2000) Ecotourism from a conceptual perspective, an extended definition of a unique tourism form. *International Journal of Tourism Research* 2, 189–202.

Brown, S. (2005) Travelling with a purpose: Understanding the motives and benefits of volunteer vacationers. *Current Issues in Tourism* 8 (6), 479–496.

Brown, S.L., Nesse, R.M., Vinokur, A.D. and Smith, D.M. (2003) Providing social support may be more beneficial than receiving it: Results from a prospective study of mortality. *Psychological Science* 14, 320–327.

Bykov, A. (2016) Altruism: New perspectives of research on a classical theme in sociology of morality. *Current Sociology* 65, 797–813.

Callanan, M. and Thomas, S. (2005) Volunteer tourism: Deconstructing volunteer activities within a dynamic environment. In M. Novelli (ed.) *Niche Tourism: Contemporary Issues, Trends and Cases* (pp. 183–200). Oxford: Elsevier Butterworth-Heinemann.

Cialdini, R.B., Schaller, M., Houlihan, D., Arps, K., Fultz, J. and Beaman, A.L. (1987) Empathy-based helping: Is it selflessly or selfishly motivated? *Journal of Personality and Social Psychology* 52, 749–758.

Clavien, C. and Chapuisat, M. (2013) Altruism across disciplines: One word, multiple meanings. *Biology & Philosophy* 28, 125–140.

Cnaan, R.A., Handy, F. and Wadsworth, M. (1996) Defining who is a volunteer: Conceptual and empirical considerations. *Nonprofit and Voluntary Sector Quarterly* 25, 364–383.

Coghlan, A. (2015) Prosocial behaviour in volunteer tourism. *Annals of Tourism Research* 55, 46–60.

Coghlan, A. and Fennell, D. (2009) Myth or substance: An examination of altruism as the basis of volunteer tourism. *Annals of Leisure Research* 12, 377–402.

Coghlan, A. and Weiler, B. (2018) Examining transformative processes in volunteer tourism. *Current Issues in Tourism* 21, 567–582.

Cohen, E. (1987) 'Alternative tourism'—A critique. *Tourism Recreation Research* 12, 13–18.

Courville, S. and Piper, N. (2004) Harnessing hope through NGO activism. *The Annals of the American Academy of Political and Social Science* 592, 39–61.

Curry, O.S., Rowland, L.A., Van Lissa, C.J., Zlotowitz, S., Mcalaney, J. and Whitehouse, H. (2018) Happy to help? A systematic review and meta-analysis of the effects of performing acts of kindness on the well-being of the actor. *Journal of Experimental Social Psychology* 76, 320–329.

Daldeniz, B. and Hampton, M.P. (2010) VOLUNtourists versus volunTOURISTS: A true dichotomy or merely a differing perception? In A.M. Benson (ed.) *Volunteer Tourism: Theoretical Frameworks and Practical Applications* (pp. 30–42). Abingdon: Routledge.

Del Pozo, E. (2002) A sociological approach to the study of altruism: The case of Amnesty International. *Michigan Sociological Review* 16, 39–62.

Dovidio, J.F., Piliavin, J.A., Schroeder, D.A. and Penner, L.A. (2017) *The Social Psychology of Prosocial Behavior*. New York: Psychology Press.

Fennell, D.A. (2006) Evolution in tourism: The theory of reciprocal altruism and tourist–host interactions. *Current Issues in Tourism* 9 (2), 105–124.

Fennell, D.A. (2008) Responsible tourism: A Kierkegaardian interpretation. *Tourism Recreation Research* 33, 3–12.

Genda, Y. (2016) An international comparison of hope and happiness in Japan, the UK, and the US. *Social Science Japan Journal* 19, 153–172.

Gray, N.J. and Campbell, L.M. (2007) A decommodified experience? Exploring aesthetic, economic and ethical values for volunteer ecotourism in Costa Rica. *Journal of Sustainable Tourism* 15 (5), 463–482.

Guttentag, D.A. (2009) The possible negative impacts of volunteer tourism. *International Journal of Tourism Research* 11, 537–551.

Hamilton, W.D. (1964) The genetical evolution of social behaviour II. *Journal of Theoretical Biology* 7, 17–52.

Haski-Leventhal, D. (2009) Altruism and volunteerism: The perceptions of altruism in four disciplines and their impact on the study of volunteerism. *Journal for the Theory of Social Behaviour* 39, 271–299.

Hoffman, M.L. (1978) Psychological and biological perspectives on altruism. *International Journal of Behavioral Development* 1, 323–339.

Hoffman, M.L. (1981) Is altruism part of human nature? *Journal of Personality and Social Psychology* 40, 121.

Holmes, K., Lockstone, L., Smith, K. and Baum, T. (2007) Volunteers and volunteering in tourism: Social science perspectives. In I. McDonnell, S. Grabowski and R. March (eds) *CAUTHE 2007: Tourism-Past Achievements, Future Challenges* (pp. 1024). University of Technology Sydney, Sydney. Conference proceedings.

Jacob, C. and Guéguen, N. (2012) Exposition to altruism quotes and helping behavior: A field experiment on tipping in a restaurant. *Annals of Tourism Research* 39, 1694–1698.

Kirillova, K., Lehto, X. and Cai, L. (2015) Volunteer tourism and intercultural sensitivity: The role of interaction with host communities. *Journal of Travel & Tourism Marketing* 32, 382–400.

Knollenberg, W., McGehee, N.G., Boley, B.B. and Clemmons, D. (2014) Motivation-based transformative learning and potential volunteer tourists: Facilitating more sustainable outcomes. *Journal of Sustainable Tourism* 22, 922–941.

Knox, T.M. (1999) The volunteer's folly and socio-economic man: Some thoughts on altruism, rationality, and community. *The Journal of Socio-Economics* 28, 475–492.

Krebs, D.L. and Van Hesteren, F. (1994) The development of altruism: Toward an integrative model. *Developmental Review* 14, 103–158.

Krippendorf, J. (1987) *The Holiday Makers: Understanding the Impact of Leisure and Travel*. London: Heinemann.

Lai, F.H., Siu, A.M., Chan, C.C. and Shek, D.T. (2012) Measurement of prosocial reasoning among Chinese adolescents. *The Scientific World Journal* 2012, article number 174845.

Lee, H.Y. (2020a) Do the locals really feel good? Understanding wellbeing in volunteer tourism from the perspectives of host communities in Mongolia. *Journal of Tourism and Cultural Change* 19, 1–26.

Lee, H.Y. (2020b) Understanding community attitudes towards volunteer tourism. *Tourism Recreation Research* 45, 445–458.

Lee, S. and Yen, C.-L. (2015) Volunteer tourists' motivation change and intended participation. *Asia Pacific Journal of Tourism Research* 20, 359–377.

Lueck, M.A. (2007) Hope for a cause as cause for hope: The need for hope in environmental sociology. *The American Sociologist* 38, 250–261.

Maoz, D. (2006) The mutual gaze. *Annals of Tourism Research* 33, 221–239.

Mayr, E. (1988) *Toward a New Philosophy of Biology: Observations of an Evolutionist*. Cambridge, MA: Harvard University Press.

McGeer, V. (2004) The art of good hope. *The ANNALS of the American Academy of Political and Social Science* 592, 100–127.

McIntosh, A.J. and Zahra, A. (2008a) Journeys for experience: The experiences of volunteer tourists in an indigenous community in a developed nation-a case study of New Zealand. In K.D. Lyons and S. Wearing (eds) *Journeys of Discovery in Volunteer Tourism: International Case Study Perspectives* (pp. 166–181). Wallingford: CABI.

McIntosh, A.J. and Zahra, A. (2008b) *Journeys for Experience: The Experiences of Volunteer Tourists in an Indigenous Community in a Developed Nation – A Case Study of New Zealand*. Wallingford: CABI.

Monroe, K.R. (1998) *The Heart of Altruism: Perceptions of a Common Humanity*. Princeton, NJ: Princeton University Press.

Mustonen, P. (2006) Volunteer tourism: Postmodern pilgrimage? *Journal of Tourism and Cultural Change* 3 (3), 160–177.

Ooi, N. and Laing, J.H. (2010) Backpacker tourism: Sustainable and purposeful? Investigating the overlap between backpacker tourism and volunteer tourism motivations. *Journal of Sustainable Tourism* 18, 191–206.

Pan, T.-J. (2012) Motivations of volunteer overseas and what have we learned – The experience of Taiwanese students. *Tourism Management* 33, 1493–1501.

Paraskevaidis, P. and Andriotis, K. (2017) Altruism in tourism: Social Exchange Theory vs Altruistic Surplus Phenomenon in host volunteering. *Annals of Tourism Research* 62, 26–37.

Piliavin, J.A. and Charng, H.-W. (1990) Altruism: A review of recent theory and research. *Annual Review of Sociology* 16, 27–65.

Pope, E. (2018) Tourism and wellbeing: Transforming people and places. *International Journal of Spa and Wellness* 1, 69–81.

Reisinger, Y. and Dimanche, F. (2010) *International Tourism*. Abingdon: Routledge.

Reisinger, Y. and Turner, L. (1998) Cultural differences between Mandarin-speaking tourists and Australian hosts and their impact on cross-cultural tourist-host interaction. *Journal of Business Research* 42, 175–187.

Sin, H.L. (2009) Volunteer tourism—'Involve me and I will learn'? *Annals of Tourism Research* 36, 480–501.

Sin, H.L. (2010) Who are we responsible to? Locals' tales of volunteer tourism. *Geoforum* 41, 983–992.

Singh, T.V. (2002) Altruistic tourism: Another shade of sustainable tourism: The case of Kanda community. *Tourism (Zagreb)* 50, 361–370.

Singh, T.V. (2004) *New Horizons in Tourism: Strange Experiences and Stranger Practices*. Wallingford: CABI.

Smith, M.K. and Diekmann, A. (2017) Tourism and wellbeing. *Annals of Tourism Research* 66, 1–13.

Söderman, N. and Snead, S.L. (2008) Opening the gap: The motivation of gap year travellers to volunteer in Latin America. In K.D. Lyons and S. Wearing (eds) *Journeys of Discovery in Volunteer Tourism: International Case Study Perspectives* (pp. 118–129). Wallingford: CABI.

Staats, S.R. and Stassen, M.A. (1985) Hope: An affective cognition. *Social Indicators Research* 17, 235–242.

Taylor, J. and Turner, R.J. (2001) A longitudinal study of the role and significance of mattering to others for depressive symptoms. *Journal of Health and Social Behavior* 42, 310–325.

Tomazos, K. and Butler, R. (2009) Volunteer tourism: The new ecotourism? *Anatolia* 20, 196–211.

Tomazos, K. and Cooper, W. (2012) Volunteer tourism: At the crossroads of commercialisation and service? *Current Issues in Tourism* 15, 405–423.

Trivers, R.L. (1971) The evolution of reciprocal altruism. *The Quarterly Review of Biology* 46, 35–57.

Uriely, N., Schwartz, Z., Cohen, E. and Reichel, A. (2002) Rescuing hikers in Israel's deserts: Community altruism or an extension of adventure tourism? *Journal of Leisure Research* 34, 25–36.

Uriely, N., Reichel, A. and Ron, A. (2003) Volunteering in tourism: Additional thinking. *Tourism Recreation Research* 28, 57–62.

Wearing, S. (2001) *Volunteer Tourism: Experiences That Make a Difference*. Wallingford: CABI.

Wearing, S. and McGehee, N.G. (2013) Volunteer tourism: A review. *Tourism Management* 38, 120–130.

Wentzel, K.R. (1998) Social relationships and motivation in middle school: The role of parents, teachers, and peers. *Journal of Educational Psychology* 90, 202–209.

Witvliet, C.V., Richie, F.J., Root Luna, L.M. and Van Tongeren, D.R. (2019) Gratitude predicts hope and happiness: A two-study assessment of traits and states. *The Journal of Positive Psychology* 14, 271–282.

Woosnam, K.M. and Norman, W.C. (2010) Measuring residents' emotional solidarity with tourists: Scale development of Durkheim's theoretical constructs. *Journal of Travel Research* 49, 365–380.

Wright, H. (2013) Volunteer tourism and its (mis)perceptions: A comparative analysis of tourist/host perceptions. *Tourism and Hospitality Research* 13, 239–250.

Zahn-Waxler, C., Radke-Yarrow, M., Wagner, E. and Chapman, M. (1992) Development of concern for others. *Developmental Psychology* 28, 126.

Zahra, A. (2010) Volunteer tourism as a life-changing experience. In A.M. Benson (ed.) *Volunteer Tourism: Theoretical Frameworks and Practical Applications* (pp. 90–101). Abingdon: Routledge.

Zahra, A. and McGehee, N.G. (2013) Volunteer tourism: A host community capital perspective. *Annals of Tourism Research* 42, 22–45.

Zahra, A. and McIntosh, A.J. (2007) Volunteer tourism: Evidence of cathartic tourist experiences. *Tourism Recreation Research* 32, 115–119.

Part 2

Destinations, Settings and Populations

5 Familiar Tourists as a Source for Hope, Happiness and the Good Life: *In Situ* Tourist Tales

David Bowen and Jackie Clarke

Introduction

This chapter is based on research that explores through *in situ* tourist tales how familiar places emerge for what we call familiar tourists, the critical touristic experience of such tourists in familiar places, and the resultant implications for themselves, others and destination development. With the call for contributions to this book we reflected on whether the experience of familiar tourists is usefully depicted as a source of hope, happiness and the good life. The relevance of our exploration and reflection in the aftermath of the COVID-19 pandemic – for many the antithesis of hope, happiness and the good life – is also deliberated.

Ten years ago, Philip L. Pearce (2012) recognised a neglected category of tourists whose motivation is to visit home and familiar places (VHFP) and stated that there is a need to study connected theoretical and practical themes. Pearce (2012) offers a highly stimulating synopsis but he is not entirely consistent with regard to terminology. We choose to label VHFP tourists from the outset under the umbrella term 'familiar tourists'. The places they visit are identified as familiar places and their particular tourism is identified as familiar tourism. The adjective *familiar* refers by dictionary definitions to something well-known and the noun *familiarity* refers to acquaintance, understanding and grasp. Academic discussion of familiarity extends back to Cohen (1972) and his comments on the comfort of the familiar tourist bubble. Baloglu (2001) defines three types of familiarity: informational (the extent of information used), experiential (previous experiences) and self-rated (how familiar with a place people believe themselves to be). Prentice (2004) later expanded those types from three to seven (informational, experiential, proximate, self-described, educational, self-assured and expected) and it can now be stated that familiarity is a multidimensional concept with more depth than was originally envisaged. In their review, Tan and Wu (2016) state that related concepts, specifically awareness, knowledge, experience and expertise, are often used synonymously with familiarity.

In practice, as will emerge later in the chapter, a familiar place is often revisited multiple times, with visits sometimes extending over a lifetime. However, a distinction needs to be drawn at the outset between research on familiar tourists and related research on repeat travel. Not all repeat tourists are familiar tourists. As Schofield and Fallon (2012) remark, repeat tourists are not a homogeneous group. Pearce (2012) comments on a number of discriminators that distinguish comparatively well-established research related to repeat travel and that on familiar tourists. The focus of research on repeat tourists is normally at a macro level, on a national or regional scale, using aggregate statistics. Destination reports that are practice-orientated invariably carry statistics on repeat tourists, often in contrast to first-time tourists. They typically compare elements such as expenditure and time spent; use of accommodation, attractions, and transport types; and judgements such as intention to revisit. For example, repeat travel is often connected to notions of tourist satisfaction, loyalty, and value from a single visit experience. However, as will become evident from our *in situ* tales, repeat visits of familiar tourists to a familiar place relate to a much wider conceptual range and the familiar refers to both place and the activity within that place.

A distinction also needs to be drawn in this introduction between familiar tourism and visiting friends and relatives (VFR). VFR has a narrower focus. Some VFR can be categorised within familiar tourism – for example, when visits are made to where familiar tourists grew up and where their family and childhood friends may still live. However, a consideration of existing literature shows that much discussion of VFR is only applicable in part to familiar tourism or not applicable at all (Backer, 2012; Backer *et al.*, 2017; Uriely, 2010). VFR is not synonymous with familiar tourism.

There is much literature relevant to a study of familiar tourism beyond that highlighted above (familiarity, repeat tourism and VFR tourism). In broad terms studies around the nexus of place and tourist behaviour are the most relevant streams. Pearce (2012: 1025) described familiar tourism as a 'messy and multi-faceted' phenomenon with a suggestion of many overlapping literature streams. Those can be subdivided to include the following: space and place (Cresswell, 2014; Relph, 1976; Tilley, 2006; Tuan, 1977); place attachment (Hamitt *et al.*, 2006; Lewicka, 2011; Low & Altman, 1992; Scannell & Gifford 2017; Williams, 2013); place-making (Dupre, 2018; Iaquinto, 2020; Lew, 2017); cocreation (Lusch & Vargo, 2006; Prebensen *et al.*, 2013; Richards, 2020); memorable experiences (Kim *et al.*, 2012; Tung & Brent Ritchie, 2011), including tourist transformation (Filep & Laing, 2019; Kirillova *et al.*,

2017) and well-being (Knobloch *et al.*, 2017). It is not the platform here to launch into a detailed literature review of such an indicative selection of authors. Rather the emphasis is on a holistic empirical exploration of the connection of familiar tourists to their familiar place, with the identification of how familiar places emerge for familiar tourists, the critical touristic experiences that characterise their relationship with their familiar place and the resultant potential implications for themselves, others and destination development.

Our initial foray into research on familiar tourists involved two preparatory focus groups, which in turn informed the critical core of our primary field research: specifically, 51 face-to-face, semi-structured *in situ* interviews with a total of 108 familiar tourists over a seven-month period. To affirm our understanding of context and ensure credence, so that what we present is recognised and understood, a spectrum of tourism providers was also engaged in conversation prior to, during, and after the stage of tourist interviews.

The tourist interviews were conducted in two case study areas, Gower and Mawddach, within Wales, UK. Both are rural and comparatively peripheral within the UK space, and both are long-established tourist destinations, albeit with different geographic markets. Gower is designated as an Area of Outstanding Natural Beauty for planning purposes and Mawddach is contained within the Snowdonia National Park.

The following three research questions helped guide us to our reflective evaluation of familiar tourists as a source for hope, happiness and the good life:

- How do familiar places emerge for familiar tourists?
- What are the critical touristic experiences of familiar tourists in familiar places?
- What are the potential implications for themselves, others and destination development?

With regard to emergence, the focus is on the locational origin of the familiar tourist–familiar place nexus and a more distinct categorisation of places beyond the home area. Regarding critical touristic experience, the focus is on identifying experiences that set familiar tourists and their tourism apart from other related tourisms (including repeat tourism). That will extend much tourism study that otherwise remains confined in large part to a view of tourism as centred on the non-familiar. Finally, with regard to implications, the focus is on the individual familiar tourist and their family, friends and contacts, plus a projection to the implications for destination development.

It is apparent that some tourists visit places they are familiar with. However, how those places become familiar to them, their experience in those places, and the potential implications deserve more scholarly empirical attention. What will emerge is that the *in situ* tales of familiar tourists are not yet captured in either their wholeness or nuance, so that the contribution of familiar tourism, including its role as a pathway to hope, happiness and the good life, is frequently overlooked.

In Situ Field Research

The build-up to *in situ* field research in Gower and Mawddach involved several stages. The catalyst from the gap identified by Pearce (2012) and the subsequent literature review led to an early preparatory exploration of ideas within a focus group setting. Two focus groups were conducted with residents in the UK city where the authors live, involving volunteer participants who identified themselves as familiar tourists. There was a total of 13 participants (5 male and 8 female) ranging in age from 26 to 75 and covering six nationalities. Familiar places included world cities (e.g. Paris, France), smaller towns and cities (e.g. Lurgan, Northern Ireland), rural areas (e.g. Lake District, England) and islands and island groups (e.g. Skye, Scotland). The focus groups were organised around a structured schedule of themes and a word association task. Informants introduced their familiar place(s) and shared the duration of such familiarity along with visit frequency; what they thought, felt and did in their familiar place(s); the meaning that familiar place(s) held for them when present or absent; any negative aspects of their relationship with familiar place(s); and how they thought the familiar place(s) gained from their visit. The focus groups were recorded, and the resulting transcripts were analysed in tandem by the two authors, leading to the joint construction of a mind map. The map informed the construction of a template for face-to-face, semi-structured *in situ* interviews with familiar tourists in Gower and Mawddach. The *in situ* tourist interviews formed the core of field research and they were in turn transcribed and analysed in tandem by the authors to produce an evidence-based evaluation of familiar tourism.

The research was exploratory. We very deliberately researched *in situ* and not in a setting away from where the experience was occurring, such as an office room or pre-arranged location divorced from ongoing tourist activity. After approaching a potential interviewee, we briefly introduced our research and followed up with an initial question that drew the interviewees into the research: 'Is Gower (Mawddach) a "familiar place" for you and, if so, to what extent?' If the interviewees answered that Gower (Mawddach) was not a familiar place, they were thanked, and no interview was conducted. The great majority of potential interviewees had no problem with the terminology of the opening question. Comparatively few interviewees asked for more information. As the work was exploratory, a formal opening definition would have overly narrowed down the exploration. There is discussion on the relevance of self-rated familiarity in Prentice (2004) and an argument is made that self-rated familiarity can be confused with experiential familiarity. However, that was not the case in this field research, most probably because the research was conducted *in situ*. In the terminology of Prentice, interviewees who self-rated themselves as familiar were also very able to self-describe their familiarity.

Part of the later questioning in the interviews revolved around whether the interviewees had other familiar places in which they had the same sort of experience and how they organised those other familiar places in relation to Gower (such as Mawddach). During that questioning it was again apparent that the interviewees were able to clearly distinguish between familiar places and non-familiar places and between their experience as familiar and non-familiar tourists. From such evidence we can state with some confidence that the interviews produced valid data that was subsequently shaped into a valid evidence-based evaluation. The self-rated familiarity with place was always backed up in what the interviewees stated during the interviews. Following self-selection, the remainder of the interview teased out the details of the interviewees' experience of the place.

Two case study areas of Gower and Mawddach were chosen because they provide sufficient similarity to add depth and sufficient difference to add nuance. They are similar inasmuch as they are both peripheral areas in Wales; both subject to planning control as a result of Area of Outstanding Natural Beauty and National Park designation, respectively; and both have a heritage as tourist destinations. Additionally, they have both been subject to some economic stress as a result of decline in farming, often said to be a stimulant for rural tourism growth (Oriade & Robinson, 2017). The role of tourism in the areas has also been subject to debate, as witnessed by a series of public meetings held by the Institute of Welsh Affairs in Gower during the early preparatory phase of the research, 2013–2015, as well as the ongoing pressure group activity of The Gower Society (2020). The case study areas are different insomuch as they are in South Wales and Mid Wales, respectively, which have contrasting physical and cultural backgrounds. Ancillary knowledge of tourism within the two field study areas was provided through contact with tourism providers (prior to, during and after the tourist interviews) and a long association of one author with Gower and the other with Mawddach.

The *in situ* interviews were conducted in a range of locations within Gower and Mawddach at different times of the day and in both high season and shoulder season. In Gower interviews were conducted at six beach, cliff and settlement locations on the north, south and west of its peninsula. Overall, 30 interviews were conducted with a total number of 67 familiar tourists. There was an even split of male and female respondents (34 female, 33 male) and a range across the age groups (18–30 (20), 31–40 (9), 41–50 (3), 51–60 (11), 61–70 (18), 71+ (6)). Nearly all the interviewees were resident in the UK, with the majority from regions stretched along the traditional source area that lies along the M4 motorway corridor from Swansea through to Cardiff, Bristol and London, 150 miles to the east. In Mawddach, 21 interviews were conducted with a total number of 41 familiar tourists in seven locations: two beaches, a spit point, coastal settlement, inland estuary bridge, inland lakes and an inland campsite. The demographic split was comparable to Gower. The Mawddach's traditional source areas were well represented: the English Midlands and English North West, 75–100 miles away. Overall, in both locations (Gower and Mawddach) 51 separate interviews were conducted with 108 familiar tourists. Interviews with two or more tourists were highly informative. Tourists in family or friendship pairs, or occasionally threes, were comfortable in each other's company, clearly thought independently and often generated rich, fast-paced discussion.

There was a determined intention to engage in relevant conversation with tourists *in situ*, whilst holidaymaking, and to explore what they thought, felt and did as familiar tourists via a qualitative study. Accordingly, data on items such as employment or income range, as commonly sought in quantitative surveys for statistical analysis of sub-groups, was not requested. However, interviewees revealed much evidence about themselves as part of their narrative. For example, they talked about their familiar place as a way to release stress from their jobs and so the breadth of trades and professions of the interviewees became clear. Among those employed there was a hairdresser, nurse and university lecturer. Among retirees there was a miner, high school teacher and architect. The questions were not asked or answered in a staccato or rigid manner, and it is judged that the conversational style allowed the creation of a rich interview experience and set of transcriptions. In most cases interviewees talked freely about sometimes sensitive matters (the role of the familiar place at times of difficulty – e.g. family problems and so on).

From the above it can be observed that all interviews were in a public or commercial space *in situ*, most often within an outdoor physical setting where the tourists were engaged in tourist activity. Overall, most interviews lasted upwards of 20 minutes, although several interviews lasted for over 30 minutes. It was striking just how much information and how many ideas were gleaned from the interviews. That is most probably related to the *in situ* location, which meant that interviewees were easily propelled into the heart of the questioning.

An interview was typically initiated after a 15–30-minute gap from the preceding interview in order to write quick field notes (for example, descriptions of interviewees' appearance or specific weather conditions during an interview) to help future recall of interviews, check equipment, and so on. In some more secluded locations, the interviewers sometimes delayed initiating an interview when it was observed that a tourist appeared in a particularly reflective mode. Breaking in on apparent deep reflection was considered insensitive.

In accordance with ethical guidelines, the broad nature of the research was outlined at the start of each interview, the interviewers did not ask for any name or contact details, and the interviewees were told that all quotes would be non-attributable and that they could discontinue the interview at any time.

The series of interviews were continued beyond the point at which it was thought that data saturation was reached when very limited extra details and ideas were generated. Analysis used an adapted version of classic qualitative research in order to break down and put back together transcribed data into higher level themes with a clear analytic story (Saldaña, 2013; Silverman, 1993; Strauss & Corbin, 1990). In the first instance recordings were listened to separately on multiple occasions by each of the two authors. That process was repeated once transcripts were produced in order to develop full familiarity with the data (Braun & Clarke, 2006) through an iterative, cyclical process. Emphasis was placed on the search for *in vivo* codes via a line-by-line analysis of the transcripts. In such a way the authors kept close to the data and alive to what was emerging, constantly open to new things, even though they carried forward some *a priori* structuring of codes into the analytic process from reading of existing theory. Empirical observation was dominant over theoretical deduction. As an example of *in vivo* coding, the word 'love' as referring to place was frequently revealed in the transcripts, as was the word 'death' in reference to interviewees' wish, in numerous cases, to have their ashes spread in a sub-location within their familiar place. Following separate interrogation of the data, the authors increasingly used team discussions to examine the transcripts and cross-compare for patterns at the unit level of both interviews and fieldwork areas (Gower and Mawddach). In so doing this higher order codes were developed: an interpretative representation of what interviewees were saying. In the team discussions use was made of a variety of techniques, including visualisations, as encouraged by D.G. Pearce (2012). A final example of visualisation (Figure 5.1) will be presented later in the findings and discussion. Figure 5.1 illustrates the outcome of the analytic process – for example, the theorisation of three overarching, higher order themes interpreting the experience of familiar tourists ('unfamiliar in the familiar', 'unexpected in the expected' and 'emotional charge'). Through adherence to sound methodological practice, as detailed earlier, the results presented in this chapter meet the various criteria for quality in qualitative research (Lincoln & Guba, 1985; Tracy, 2010).

In Situ Tales

The *in situ* voices of familiar tourists as documented during visits to their familiar places are grouped into three discrete but connected behavioural parts: emergence, experience and implications. Each has a recognisable cluster of components. The behaviours of familiar tourists also suggest implications for destination development as a familiar place. A visualisation is provided in Figure 5.1. The voices of the *in situ* interviewees are expressed in a series of quotes, all of which are from different interviewees. Interviewees are identified by gender (F/M), age, and location – for example, M 40, Mawddach. There are two instances in which the gender, age, and location of different interviewees are the same (F 55, Gower and F 65, Gower) and in those instances a further sub-division (i, ii) has been inserted.

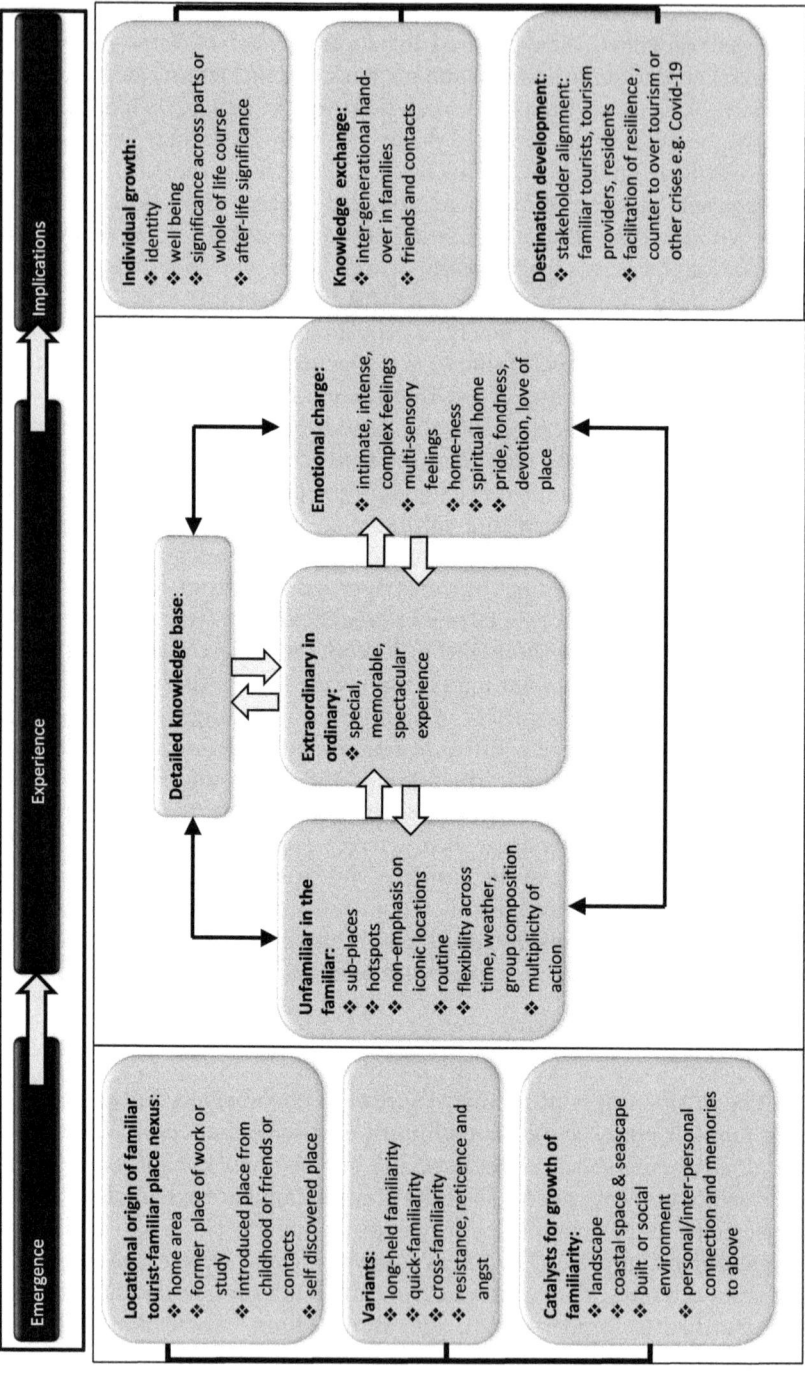

Figure 5.1 Familiar tourist-familiar place nexus

Emergence

From the interviews we are able to extend and specify the locational origin of a familiar place to include the home area in which tourists grew up, other places that tourists have relationships with from work or study, introduced places from childhood holidays or holidays in adult life with friends or contacts, and also self-discovered holiday places (Figure 5.1). This fourfold division of locational origin has not previously been categorised. Pearce (2012) was not consistent in discussion of home and familiar place locations and in any case produced a conceptual listing rather than one based on empirical evidence. Variants of familiarity (long-held, quick and cross) and counters to familiarity (resistance, reticence and angst) are also specified in Figure 5.1.

The first tale below relates to Gower as a place visited when studying at university, whilst the second tale describes Gower as an introduced place from a childhood family holiday:

F 35, Gower: Oh, well, we came here a lot while I was at university, which was quite a while ago, 15 years or something now. We used to just come for the weekend for beach parties and walking and stuff. So I've got some memories associated with it as well as this beach [still] being lovely, quiet, etc.

F 70, Gower: My parents had a static caravan. We came for five or six weeks. I remember having *wonderful* summers with friends and just having freedom to roam around the dunes and just having a great time.

There are variants of familiarity: long-held familiarity developed over many years; quick familiarity developed from very few visits; and cross-familiarity developed from comparison with other places of perceived similarity (Figure 5.1). Here is an example that is a hybrid of quick and cross-familiarity:

Interviewer: You said that you had been here about three times.
F 28, Gower: Yeah, I'd say three or four times, yeah.
Interviewer: OK, so what got you here originally?
F 28, Gower: I'd say people I know. But your earlier question [was] 'What brings me back?' There are obviously other places to explore ... [but] I love it here. ... It's beautiful; it reminds me a lot of home. I'm from South Africa and here, like there, you know, there are big beaches, lots of sand, good waves, warm water, a lot of green spaces ...

Cross-familiarity has similarity to Scannell and Gifford's (2013) description of place attachment to a specific class of place or generic type (e.g. big beaches and wild rivers). A non-hybrid example of cross-familiarity is evident from a Canadian artist who partly attributed her established love of the Mawddach to her attachment for the wild rivers of Canada:

F 32, Mawddach: I love rivers and I especially like wild rivers and Mawddach, although it is not as wild as some of the rivers in Canada. Obviously, it is not. I mean London [*where she now lives with her English husband and son*], the Thames, is brilliant because it is a fascinating working river, but this is a proper wild river to me and I really like that. So it is geography as well, isn't it, that draws us.

The growth of familiarity is not a straight-line trajectory. Tourists can sometimes take a break from their familiar place and on occasion they develop resistance, reticence and angst about returning to a familiar place because they feel judged by others on their desire to return rather than explore a novel place. The catalysts for growth of familiarity embrace aspects of the physical environment (especially landscape and seascape in the field destinations) and the built or social environment, all ranged alongside personal or interpersonal experience, connection and resultant memories (Figure 5.1). These are all themes that are present as catalysts in discussion of the literature on tourists, place and space, place attachment and place-making. The familiar place of most tourists is imbued with a sense of place (Relph, 1976; Tuan, 1977) and the tourists show strong place attachment (Hamitt *et al.*, 2006). However, while familiar places are consumed in the normal tourist sense, the importance of the catalysts, compounded time and again on subsequent visits, is very much stressed by familiar tourists and resonates with the bottom-up, organic processes of place-making identified by Lew (2017). One group of family and friends recalled, reflectively and excitedly, the way in which they marked the dawn of 2000 in.their familiar place of Rhosili Beach, Gower. The extract below illustrates growth and compounding of familiarity. There was no other place that they wanted to be on such a special night:

F 65i, Gower: We keep our [static] caravan at a farm. We took over a field for the night of the millennium and had a marquee and a four-course meal. One family member cooked and we had loads of family members who all camped in tents or caravans or something. And then at midnight we went down onto Rhosili Beach and had fireworks and music going, and I swam at midnight.

F 75, Gower: And a fire. It was brilliant, absolutely brilliant. And then in the morning, it was gorgeous weather, and we sat having a champagne and salmon and scrambled egg breakfast out in the open, in a semi-circle, because it was so warm and lovely. Yes and it felt, seemed, an important place to see the millennium in.

Such tales about the emergence of familiarity lead us from the first to the second discrete but connected behavioural part: experience.

Experience

Overall, three critical touristic experiences of familiar tourists are worthy of emphasis (Figure 5.1), all rooted in a very detailed knowledge base. First, familiar tourists can make full use of space to search for the unfamiliar in the familiar. In that sense they seek novelty (Mitas & Bastiaansen, 2018) but in a familiar rather than unfamiliar place. As a further twist, the familiar place often pivots around specified named sub-places or personal hotspots. Familiar tourists are not confined to iconic locations, as identified, too, in an exploration of familiarity and sub-destination choice among international tourists (Lee & Tussyadiah, 2012). Often it is seemingly routine activities in the sub-places (e.g. walking in familiar territory, taking photographs) that are at the forefront of tourist descriptions:

F 45, Mawddach: Every year I take photographs out there [Mawddach estuary]. Even though it is the same view it has so many moods, so there are always different lights, different clouds, tide-in, tide-out . . .

The routine activities can develop into rituals:

F 28, Mawddach: As a tradition we have to go walk up Fegla Fawr [Mawddach] and have a picnic where Granny's ashes are scattered.

Some interviewees recalled how they found it difficult when an attempt was made to break rituals, often linked to such sub-places. However, the ritual nature of such activities was not considered to be a negative thing:

F 35, Gower: You try different places and then you end up going back, you know, to the original ones.

Ehn and Löfgren (2010) argue that repetitive actions are an essential part of people's individual and social lives. However, the evidence from *in situ* interviews with familiar tourists suggests that the apparent repetition of activity does not lead to a mundane experience. Rather, through the opportunity offered by familiar tourism, tourists manage to transcend the mundane and find the unfamiliar in the familiar.

Cocreation of value and individual or group operant resources of familiar tourists, allied with involvement of effort, time and money (Holbrook, 2006; Prebensen *et al.*, 2013) allow for a multiplicity of varied actions in the familiar place. For example, in line with the thinking around value-in-use and tourists as resource integrators (Lusch & Vargo, 2006; Vargo & Lusch, 2006), each experience of the same beach is unique: a place to walk, sit, or avoid in one type of weather or in one time of day or with one set of companions. Tourists were able to casually pinpoint the names and characteristics of a range of locations and with their detailed knowledge they were able to tailor their behaviour to fit the day, even in a comparatively small geographic area:

F 55i, Gower: It's not like being a [new] visitor in that you could maybe come for a week and you've maybe not seen what you want to see. If you know where you're going . . . you know where to go, where to walk, and what to do.

As a final aside in this section, while it is clear that some tourists use the same accommodation time after time (e.g. a static caravan or second home), others deliberately return to different accommodation. The familiar place by no means requires an established base and varying the accommodation is sometimes part of the way in which tourists seek the unfamiliar in the familiar.

The second critical touristic experience of familiar tourists that derives from a very detailed knowledge base is their discovery of the extraordinary in the ordinary (Figure 5.1). They enjoy special, memorable and even spectacular experiences from everyday occurrences in their familiar place. One familiar tourist, a barrister, narrated what it felt like a day after visiting her familiar place:

F 65ii, Gower: I went into work the next day and I still had this huge smile on my face, and people sort of said, 'Goodness, why are you smiling so much all day?' And I was just remembering being in the water, the sun, and everybody else enjoying coming in on the surf and all that sort of stuff. . . . I thought about it [the familiar place] when I was not there, especially when working. You know, when really stressed or something. . . . There were moments, when I was being given a really hard time, when I'd just say to myself, 'Just think of Mewslade (Bay) and just think of swimming, and this [work stress] is going to be over in a minute.' And, you know, I'd just put myself there and that would just, sort of, calm me down.

Cohen and Cohen (2012) comment on the end of binary comparisons. That is certainly the case from the evidence of how familiar tourists search for the unfamiliar in the familiar and the extraordinary in the ordinary. Contrary to what might seem the case, the activity patterns of familiar tourists characterise them as one particular sort of hybrid tourist (Boztug *et al.*, 2015).

The detailed knowledge base of familiar places also generates a distinctive emotional charge (Figure 5.1). Familiar tourists have noticeably strong affective bonds (Coghlan *et al.*, 2012) with their familiar place and emphasise the affective over the cognitive. The very first interview in Gower was with a London hairdresser who had rediscovered a part of a beach that she remembered and connected with from visits as a child. An extract from her interview is intimate, intense, and complex. It is multi-sensory, and it is also an experience that is embodied (Obrador-Pons, 2003), not just involving the tourist gaze, like the previous quote from the barrister:

F 55ii, Gower: Yeah, well it's part of my childhood. It's one of my very, *very* sweetest memories of my childhood: this time of year, with my family. I *absolutely* loved it here, and I remember having many, *many* happy times playing on this beach, exactly here. There was hardly anyone here [then] and today it's the same: it's beautiful, it's Monday, it's mega. Really, really, and it's very special to me. On the walk down through the dunes, you see all the really different types of flora and fauna and different wild plants. You know, I'm from London now, I mean we've lived there for years, and it's just *so* nice to see this. It's been heaven!

The emotions of familiar tourists lead to the seduction of place. Familiar tourists often view their familiar place as a place that has the rootedness and at-home-ness characteristic of their non-tourist home (Relph, 1976; Windsong, 2010). Moreover, it is not uncommon for familiar tourists to consider their familiar place as a spiritual home. They speak of pride, fondness, devotion, and love attached to their familiar place. Typical illustrations, here from Mawddach, include 'I fell in love' (F 72) and 'It definitely captured my heart' (F 47), demonstrating strong place attachment.

With such details on emergence and experience we are able to highlight implications as they apply to tourists and others, and also destination development.

Implications: Tourists and Others

With regard to individual growth (Figure 5.1), familiar tourists repeatedly articulate how the emergence of a familiar place, and the resultant experience can enhance a sense of personal and social identity and deliver a sense of well-being. Oftentimes such transformation was the motive for return, in contrast to the unforeseen trigger to transformation described by Kirillova *et al.* (2017). Moreover, familiar tourism can retain significance over a whole life course.

The experiences of the hairdresser (F 55ii, Gower) and the barrister (F 65ii, Gower) quoted above illustrate individual and social identity that derives from their visits to familiar places and how personal well-being has been added to their lives from visits to their familiar places, in their cases across a whole life course. All such outcomes were clearly and frequently expressed by familiar tourists. The familiar tourist–familiar place link also extends its significance, not unusually, after death. As interviewers we started to anticipate when interviewees were about to mention the hold of their familiar place after death, on occasion expressed when the recording was stopped and so captured via field notes. Here is an example from an interviewee who shared some tea with an interviewer after finishing kayaking for the day:

M 36, Mawddach: I know it sounds macabre, but I'd like to have my ashes scattered here.

The earlier example (F 28, Mawddach) regarding a ritual trip to where a grandparent's ashes were scattered also illustrates what we call as interviewers an 'ashes moment'. Such revelations became signifiers of data saturation and acted as markers for establishment of rapport.

In addition to using knowledge for their own purpose it is noticeable that familiar tourists also exchange knowledge, passing on what they know about their familiar place and their enthusiasm for their familiar place to family, friends, and contacts (Figure 5.1). Parents and grandparents commonly seek to pass on knowledge and experience to children. For the most part they craft a forward-looking intergenerational handover of their familiar place that links to the emphasis of Epp and Price (2008) on intergenerational transfer as a means to strengthen collective family identity. Gitelson and Crompton (1984) include the desire to show a destination to others as a motivation for repeat travel. Little has been written on that since their study (Schofield & Fallon, 2012) but it comes through very strongly as an outcome of the experience of familiar tourism.

Implications: Destination Development

Implications extend beyond individual growth and knowledge exchange. In most tourism destinations there are three primary stakeholder groups: tourists, providers, and residents. Recognition of tourist behaviours evident from this empirical study and emphasis on familiar tourism in a destination, with alignment of stakeholders and enhancement of a common vision and endeavour, can create an opportunity for destination development that increases destination resilience (Figure 5.1). That is especially apt in a world that faces change from COVID-19, in the short term at the very least, as well as the threat of further pandemics or other crises.

Beritelli *et al.* (2014) identify the danger of non-alignment between stakeholder groups within destinations, a situation that is recognised by many other authors (Boley *et al.*, 2018; Dupre, 2018; Jamal & Getz, 1995). However, familiar tourism provides a basis for alignment of two sets of stakeholders: familiar tourists and tourism providers, on the one hand, and familiar tourists and residents, on the other.

With regard, first, to stakeholder alignment of familiar tourists and tourism providers, background for this study was developed through conversations with a range of providers from businesses and tourism-related organisations. Few providers appreciated the extent of familiar tourism or had knowledge of the behaviour of familiar tourists. Increasingly, destination development requires the need to compete through tourist experience (Mathis *et al.*, 2016). One crux of memorable experience identified by

Kim (2014) is detailed knowledge of place and that is the driver of the unfamiliar in the familiar, the extraordinary in the ordinary, and the emotional charge identified within the *in situ* interviews. Tung and Brent Ritchie (2011) comment that the delivery of memorable experiences needs to be made probable and that seems more realistic for providers to aspire towards via the behaviour patterns of familiar tourists.

With the caveat that the *in situ* interviewees in this research were dominantly domestic tourists, it seems that familiar tourists bring other major advantages for providers. On the pragmatic side, familiar tourists can more easily manipulate their visits according to circumstance: they are often not season-dependent, or weather-dependent, or so subject to other vagaries caused by economic, social or political circumstance. They are a stable market not overly subject to flux. Familiar tourists can have a calling for their tourist place that is summed up in the Welsh words *hiraeth* (longing) and *cynefin* (a place to stand), the latter concept being similar to the German *heimat* (home/homeland). Also, through their emotional charge with their familiar place, they can act as marketing foot soldiers and engage like-minded people who in turn become familiar tourists. In this behaviour, they shift position somewhere along the continuum from Lew's (2017) organic place-making towards more deliberate or planned place-making. However, destination management study rarely appreciates that tourists can engage with a destination way beyond the time limit of the actual visit (Saraniemi & Kylanen, 2011), something that familiar tourists are especially apt to do, both before and after a visit. As one interviewee stated:

M 50, Gower: The knowledge has been passed down to me [from parents]. I tell people a lot about it here. I'm quite proud of it, if I'm honest. I like to tell people to come here. I think it is good advice. I do give it a good press.

At present, familiar tourism does not have a strong industry lobby. It is stated by Shani and Uriely (2012) that VFR tourism does not happen naturally. The same applies to familiar tourism, of which VFR can be a part. Familiar tourists need to be recognised in whatever destination is their familiar place. Providers as stakeholders that understand their characteristics and behaviours, as evidenced in the *in situ* interviews, can align activity for consequent, mutual benefit. Providers can help create the unfamiliar within the familiar and the extraordinary in the ordinary or strive to maintain the emotional charge. They can work with destination management organisations so that the bond between tourists and place is not broken through inappropriate change pressures.

With regard, second, to alignment of familiar tourists and residents, in both Gower and Mawddach there were numerous comments by familiar tourists on favourable contacts with local residents, including aligned visions of place. There is a long-established pressure group The Gower Society whose membership is drawn from local people as well as what we identify as familiar tourists, both from the UK and internationally. There is much scope for joint creation of shared places (Giovanardi *et al.*, 2014; Richards, 2020; Sheller & Urry, 2004). Amsden *et al.* (2011) argue that place attachment can be used as a metric in tourism development, together with other more common metrics such as visitor satisfaction and ecological quality. Destinations that recognise and emphasise familiar tourism can fashion such an initiative, considering the shared view of familiar tourists and residents.

The special scope for tourist–provider–resident alignment that familiar tourism embodies makes it particularly relevant as a form of tourism to be recognised and encouraged. In tandem the elements of familiar tourism (Figure 5.1) that make up the familiar tourist–familiar place nexus allow tourism destinations to embrace rather than resist tourism. Butler (2017: 5) states that there is a need '[to] improve the ability of tourism destinations to withstand the effects of tourism or, in other words, to make them more resilient'. There are multiple perspectives of resilience and Berbes Blasquez and Scott (2017) refer to the need to specify resilience 'of what' and 'to what'. Familiar tourism facilitates resilience of tourism destinations to the tensions caused by a non-alignment of three primary stakeholder groups: tourists, providers and residents.

Familiar tourism seems particularly relevant to whether COVID-19 leads to a shaping (modification) of established tourist behaviours or a radical resetting (Hall *et al.*, 2020). In the short term, it seems likely that familiar places will feature heavily among the first destinations that tourists will return to after COVID-19 lockdown. Tourists will want to return to places with which they have personal/interpersonal connections, memories, and detailed knowledge. In the medium/long term, in a world of tourists and tourism that is shaped by COVID-19 rather than reset, due recognition of and emphasis on familiar tourism by providers and destination organisations can help offer a counter to re-emergent stresses of overtourism or other crises (Gonzalez *et al.*, 2018; Seraphin *et al.*, 2018). Alternatively, in a reset world, familiar tourism can still offer one among other realistic post-COVID-19 futures. The characteristics of familiar tourism evidenced within this chapter and its potential to generate an aligned response of tourists, providers and residents mean that it is particularly relevant whatever post-COVID-19 scenario emerges over the coming years.

Conclusion

This chapter explores how familiar tourism (an amended version of VHFP tourism) emerges for familiar tourists, their experience in familiar places, and the resultant implications for themselves, others and destination development. Tourist tales that emerge from *in situ* interviews among holidaymaking tourists in Wales, UK, illustrate the above. Overall, familiar tourists appear to engage in a positive way with tourism and walk the pathway of hope, happiness and the good life. Some not untypical quotes from one tourist in the preparatory focus groups encapsulate the way that the experience of familiar tourists can link with positive rather than forlorn hope (quote 1) as well as happiness and the good life both for oneself and others (quote 2).

F 40, focus group:

1: I am already super-happy, and [yet] I am still here [home]. I know I am going to have such a good time.
2: [When] you are in this place, you are so happy and feel so well [and] people around you benefit from that too, I think.

We connected with tourists *in situ* while engaged in touristic activity in their familiar place. It was important to catch the attention of each tourist interviewee and the terms 'familiar place' and 'familiar tourist' were quickly understood. Tourists *in situ* immersed themselves in an account of their familiar places. They told detailed stories about their familiar place, often covering many years (and even generations) and fully justified their self-description as being familiar with their tourist place.

A wider view is offered here than by Pearce (2012) with regard to the locational origin of a familiar tourist–familiar place nexus. It has also been shown that whilst existing strands of literature are applicable to familiar tourism, they require a twist because familiar tourists act in a distinct way. The very nature of familiar tourism, its emergence, experience and implications, mark it out as discrete. Edensor (2007) and Caruana and Crane (2011) outline how freedom is controlled by power brokers in the tourism industry. However, compared to tourists who are not familiar with a place, familiar tourists are freer of social controls (whether from social media, guidebooks, or even directive notices along a pathway). Familiar tourists, with their detailed knowledge base, appear more rather than less likely to fulfil experiential needs when compared with non-familiar tourists. They find the unfamiliar in the familiar, the extraordinary in the ordinary, and an evident emotional charge. The implications affect familiar tourists' growth as individuals and also knowledge exchange with family, friends and contacts. There are particular implications, too, for destination development.

There is scope for iterative extension and replication in other places, such as university towns, in which students develop lasting, intense memories of and connection to sub-places from experiences that are vividly retained as alumni. Pearce (2011: 137) himself writes passionately and informatively of the 'career souvenir' that he developed as a student in Oxford. Specific emphasis could also be placed on any one of the many interrelated aspects revealed from this holistic study (Figure 5.1). In addition, while interviews with tourists lie at the heart of the study and conversations with providers helped to provide ancillary knowledge, there was no formal contact (e.g. through interviews) with residents. A study of resident engagement with familiar tourists compared to other tourists can add an extra dimension to future research (Hwang *et al*., 2012; Jordan,

2015). It may be that the hope, happiness, and good life of familiar tourists are not reciprocated by residents. Our intention is to pursue this in the Gower case area as a follow-up to our initial study.

Other research techniques could also be employed. For example, we initially intended to use photographs in the field research to jog interviewees' memories of sub-places. That proved unnecessary because interviewees were very able to talk in depth about their familiar place. However, photographs would be a particularly good idea to use for research in places that have been subject to greater change. They would provide a baseline for discussion on the amount of change and the effect on familiarity. The use of space by familiar tourists vis-à-vis other tourists using the latest tracking technology (Grinberger & Shoval, 2019) would also enhance understanding.

As is the case with all exploratory research, there is a limit as regards generalisation. This research relates to two case study areas over a high summer season and two shoulder seasons in two parts of Wales, UK. We acknowledge how the planning regulations in our two case study areas create restrictions on development that are not always present elsewhere in the UK and beyond. Tourist experiences and implications may be very different if the detailed knowledge base and familiarity with place is beset by major change.

It is evident that familiar tourism offers a chance to create a competitive destination that is not subject to substitution by fashion or fad. However, familiar tourism is not staid as a tourism form. On the contrary, it is highly contemporary and happens to fit well with emerging tourism forms. These include slow tourism with its particular lifestyle motivation (Clancy, 2018; Oh *et al.*, 2016); creative tourism with the potential it offers for freer and more meaningful experiences for the tourist, plus a more equal relationship between the resident and the tourist (Richards, 2011, 2020); and transformation tourism (Reisinger, 2013; Kirillova *et al.*, 2017), in which a tourist goes one step beyond cocreation and seeks meaningful wider life transformation. Future links between research on familiar tourism and research on such tourism forms may yield very useful benefits for destination development. The relevance of familiar tourism in the aftermath of COVID-19 has already been raised.

Most probably familiar tourism in Gower and Mawddach reflects a common unrevealed inclination among tourists to seek familiar places, with positive resultant implications for themselves, others and destination development. A regular comment by tourists in this research on familiar tourism was 'We just keep coming back'. The reason is evident from the hope, happiness and the good life so ensured.

References

Amsden, B.L., Stedman, R.L. and Kruger, L.E. (2011) The creation and maintenance of sense of place in a tourism-dependent community. *Leisure Sciences* 33, 32–51.
Backer, E. (2012) VFR travel: It is underestimated. *Tourism Management* 33, 74–79.

Backer, E., Leisch, F. and Dolnicar, S. (2017) Visiting friends or relatives? *Tourism Management* 60, 56–64.
Baloglu, S. (2001) Image variations of Turkey by familiarity index. *Tourism Management* 22, 127–133.
Berbes Blasquez, M. and Scott, M. (2017) The development of resilience thinking. In R.W. Butler (ed.) *Tourism and Resilience* (pp. 9–22). Wallingford: CABI.
Beritelli, P., Bieger, T. and Laesser, C. (2014) The new frontiers of destination management: Applying variable geometry as a function-based approach. *Journal of Travel Research* 53 (4), 403–417.
Boley, B.B., Strzelecka, M. and Watson, A. (2018) Place distinctiveness, psychological empowerment, and support for tourism. *Annals of Tourism Research* 70, 137–139.
Boztug, Y., Babakhani, N., Laesser, C. and Dolnicar, S. (2015) The hybrid tourist. *Annals of Tourism Research* 54, 190–203.
Braun, V. and Clarke, V. (2006) Using thematic analysis in psychology. *Qualitative Research in Psychology* 3 (2), 77–101. https://doi.org/10.1191/1478088706qp063oa.
Butler, R.W. (ed.) (2017) *Tourism and Resilience*. Wallingford: CABI.
Caruana, R. and Crane, A. (2011) Getting away from it all: Exploring freedom in tourism. *Annals of Tourism Research* 38 (4), 1495–1515.
Clancy, M. (ed.) (2018) *Slow Tourism, Food and Cities: Pace and the Search for the Good Life*. London: Routledge.
Coghlan, A., Buckley, R. and Weaver, D. (2012) A framework for analysing awe in tourism experiences. *Annals of Tourism Research* 39 (3), 1710–1714.
Cohen, E. (1972) Toward a sociology of international tourism. *Social Research* 39, 164–182.
Cohen, E. and Cohen, S. (2012) Current sociological theories and issues in tourism. *Annals of Tourism Research* 39 (4), 2177–2202.
Cresswell, T. (2014) *Place: A Short Introduction*. Chichester: Wiley-Blackwell.
Dupre, K. (2018) Trends and gaps in place-making in the context of urban development and tourism: 25 years of literature review. *Journal of Place Management and Development* 12 (1), 102–120.
Edensor, T. (2007) Mundane mobilities, performances and spaces of tourism. *Social and Cultural Geography of Tourism* 8 (2), 191–215.
Ehn, B. and Löfgren, O. (2010) *The Secret World of Doing Nothing*. Berkeley: University of California Press.
Epp, A.M. and Price, L.L. (2008) Family identity: A framework of identity interplay in consumption practices. *Journal of Consumer Research* 35 (1), 50–70. https://doi.org/10.1086/529535.
Filep, S. and Laing, J. (2019) Trends and directions in tourism and positive psychology. *Journal of Travel Research* 58 (3), 343–354.
Giovanardi, M., Lucarelli, A. and Decosta, P. (2014) Co-performing tourism places: The 'Pink Night' festival. *Annals of Tourism Research* 44, 102–115.
Gitelson, R.J. and Crompton, J.L. (1984) Insight into the repeat vacation phenomenon. *Annals of Tourism Research* 11, 199–217.
Gonzalez, V., Coromina, L. and Gali, N. (2018) Over-tourism: Residents' perceptions of tourism impact as an indicator of resident social carrying capacity: Case study of a Spanish heritage town. *Tourism Review* 73 (3), 277–296.
The Gower Society (2020) The Gower Society – Living Gower. See www.thegowersociety.org.uk/en (accessed February 2021).
Grinberger, A. and Shoval, N. (2019) Spatiotemporal contingencies in tourists' intradiurnal mobility patterns. *Journal of Travel Research* 58 (3), 512–530.
Hall, C.M., Scott, D. and Gössling, S. (2020) Pandemics, transformations and tourism: Be careful what you wish for. *Tourism Geographies* 22 (3), 577–598. https://doi.org/10.1080/14616688.2020.1759131.
Hamitt, W.E., Buckland E.A. and Bixler, R.D. (2006) Place bonding for recreation places: Conceptual and empirical development. *Leisure Studies* 25 (1), 17–41.

Holbrook, M.B. (2006) Consumption experience, customer value, and subjective personal introspection: An illustrative photographic essay. *Journal of Business Research* 59 (6), 714–725.

Hwang, D., Stewart W.P. and Ko, D. (2012) Community behaviour and sustainable rural tourism development. *Journal of Travel Research* 51 (3), 328–341.

Iaquinto, B.L. (2020) Understanding the place-making practices of backpackers. *Tourist Studies* 20 (3), 336–353.

Jamal, T.B. and Getz, D. (1995) Collaboration theory and community tourism planning. *Annals of Tourism Research* 22 (1), 186–204.

Jordan, E.J. (2015) Planning as a coping response to proposed planning development. *Journal of Travel Research* 54 (3), 316–328.

Kim, J.-H. (2014) The antecedents of memorable tourism experiences: The development of a scale to measure the destination attributes associated with memorable experiences. *Tourism Management* 44, 34–45.

Kim, J., Ritchie, J.R. and McCormick, B. (2012) Development of a scale to measure memorable tourism experiences. *Journal of Travel Research* 51 (1), 12–25.

Kirillova, K., Lehto, X. and Cai, L. (2017) Tourism and existential transformation: An empirical investigation. *Journal of Travel Research* 56 (5), 638–650.

Knobloch, U., Robertson, K. and Aitken, R. (2017) Experience, emotion and eudaimonia: A consideration of tourist experiences and well-being. *Journal of Travel Research* 56 (5), 651–662.

Lee, G.L. and Tussyadiah, I.P. (2012) Exploring familiarity and destination choice in international tourism. *Asia Pacific Journal of Tourism Research* 17 (2), 133–145.

Lew, A.A. (2017) Tourism planning and place making: Place-making or placemaking? *Tourism Geographies* 19 (3), 448–466.

Lewicka, M. (2011) Place attachment: How far have we come in the last forty years? *Journal of Environmental Psychology* 31 (3), 207–230

Lincoln, Y.S. and Guba, E.G. (1985) *Naturalistic Inquiry*. Beverly Hills: Sage.

Low, S.M. and Altman, I. (1992) Place attachment: A conceptual inquiry. In I. Altman and S.M. Low (eds) *Place Attachment* (pp. 1–12). New York: Plenum Press.

Lusch, R.F. and Vargo, S.L. (2006) Service-dominant logic: Reactions, reflections and refinements. *Marketing Theory* 6 (3), 281–288.

Mathis, E.F., Kim, H., Uysal, M., Sirgy, J.M. and Prebensen, N.K. (2016) The effect of co-creation experience on outcome variable. *Annals of Tourism Research* 57, 62–75.

Mitas, O. and Bastiannsen, M. (2018) Novelty: A mechanism of tourists' enjoyment. *Annals of Tourism Research* 72, 98–108.

Obrador-Pons, P. (2003) Being on holiday: Tourist dwellings, bodies and place. *Tourist Studies* 3 (1), 47–66.

Oh, H., Assaf, G. and Baloglu, S. (2016) Motivations and goals of slow tourism. *Journal of Travel Research* 55 (2), 205–219.

Oriade, A. and Robinson, P. (2017) *Rural Tourism and Enterprise: Management, Marketing and Sustainability*. Wallingford: CABI.

Pearce, D.G. (2012) *Frameworks for Tourism Research*. Wallingford: CABI.

Pearce, P.L. (2011) Career souvenirs. In P.L. Pearce (ed.) *The Study of Tourism: Foundations from Psychology* (pp. 133–154). Bingley: Emerald.

Pearce, P.L. (2012) The experience of visiting home and familiar places. *Annals of Tourism Research* 39 (2), 1024–1047.

Prebensen, N.K., Vitterso, J. and Dahl, T.I. (2013) Value co-creation significance of tourist resources. *Annals of Tourism Research* 42, 240–261.

Prentice, R. (2004) Tourist familiarity and imagery. *Annals of Tourism Research* 31 (4), 923–945.

Reisinger, Y. (ed.) (2013) *Transformational Tourism: Tourist Perspectives*. Wallingford: CABI.

Relph, E. (1976) *Place and Placelessness*. London: Pion

Richards, G. (2011) Creativity and tourism – The state of the art. *Annals of Tourism Research* 38 (4), 1225–1253.

Richards, G. (2020) Designing creative places: The role of creative tourism. *Annals of Tourism Research* 85. https://doi.org/10.1016/j.annals.2020.102922.

Saldaña, J. (2021) *The Coding Manual for Qualitative Researchers* (4th edn). London: Sage.

Saraniemi, S. and Kylanen, M. (2011) Problematizing the concept of tourism destination: An analysis of different theoretical approaches. *Journal of Travel Research* 50 (2), 133–143.

Scannell, L. and Gifford, R. (2013) Comparing the theories of interpersonal and place attachment. In L.C. Manzo and P. Devine-Wright (eds) *Place Attachment: Advances in Theory, Methods, and Applications* (pp. 23–36). Abingdon: Routledge.

Scannell, L. and Gifford, R. (2017) The experienced psychological benefits of place attachment. *Journal of Environmental Psychology* 51, 256–269.

Schofield, P. and Fallon, P. (2012) Assessing the viability of university alumni as a repeat visitor market. *Tourism Management* 33, 1373–1384.

Seraphin, H., Sheeran, P. and Pilato, M. (2018) Over-tourism and the fall of Venice as a destination. *Journal of Destination Marketing & Management* 9, 374–376.

Shani, A. and Uriely N. (2012) VFR tourism: The host experience. *Annals of Tourism Research* 39 (1), 421–440.

Sheller, M. and Urry, J. (2004) *Tourism Mobilities: Places to Play, Places in Play*. London: Routledge.

Silverman, D. (1993) *Interpreting Qualitative Data: Methods for Analysing Talk, Text, and Interaction*. London: SAGE.

Strauss, A. and Corbin, J. (1990) *Basics of Qualitative Research*. Newbury Park: SAGE.

Tan, W.-K. and Wu, C.-E. (2016) An investigation of the relationships among destination familiarity, destination image and future visit intention. *Journal of Destination Marketing and Management* 5 (3), 214–226.

Tilley, C. (2006) Identity, place, landscape and heritage. *Journal of Material Heritage* 11 (1/2), 7–32.

Tracy, S.J. (2010) Qualitative quality: Eight big-tent criteria for excellent qualitative research. *Qualitative Inquiry* 16 (10), 837–851.

Tuan, Y.-F. (1977) *Space and Place: The Perspective of Experience*. Minneapolis: University of Minnesota Press.

Tung, V.W.S. and Brent Ritchie, J.R. (2011) Exploring the essence of memorable tourism experiences. *Annals of Tourism Research* 38 (4), 1367–1386.

Uriely, N. (2010) Home and away in VFR tourism. *Annals of Tourism Research* 37 (3), 854–857.

Vargo, S.L. and Lusch, R.F. (2006) Service-dominant logic: What it is, what it is not, what it might be. In R.F. Lusch and S.L. Vargo (eds) *The Service-Dominant Logic of Marketing: Dialog, Debate and Directions* (pp. 43–56). Armonk, NY: M.E. Sharpe.

Williams, D.R. (2013) 'Beyond the commodity metaphor', revisited: Some methodological reflections on place attachment research. In L.C. Manzo and P. Devine-Wright (eds) *Place Attachment: Advances in Theory, Methods, and Applications* (pp. 89–99). Abingdon: Routledge.

Windsong, E.A. (2010) There is no place like home: Complexities in exploring home and place attachment. *The Social Science Journal* 47, 205–214.

6 Food Tourism through the Lens of Post-Materialism: Valuing the Cultural Ruralscape

Subhajit Das and Hiran Roy

Introduction

A satisfactory experience at a destination can affect tourists' emotions positively (Filep & Pearce, 2013) and may contribute to elements of positive psychology like hope, happiness and prospective life satisfaction (Sirgy, 2002). Such positive emotions further enhance their quality of life by triggering social, cultural, intellectual and physical development (Gump & Matthews, 2000). An experience is argued to be the outcome of the perception emerging from the 'body-mind exercise' (Matteucci, 2016: 55). Thus, sensory activities experienced by tourists like seeing, smelling, and tasting are of significant importance in tourism literature. Sensory activities like tasting food can trigger positive emotions for tourists, which in turn ensures self-growth, enhanced quality of life and well-being for them in the long run (Fredrickson, 2001). Host communities also benefit from such sensory activities in promoting local products, gaining economic benefits, receiving recognition for their cultural values, conserving traditions, and so on (Hillman *et al.*, 2016). While the cultural context is prominent in culinary tourism, food tourism primarily focuses on tourists' physical activities and experiences related to food consumption (Ellis *et al.*, 2018; Hall *et al.*, 2004).

Food tourism generates ample scope for developing a region holistically in its culinary product form, including rural areas (Ellis *et al.*, 2018; Hall & Gössling, 2016). Rather than promoting food as an independent tourism product, rural stakeholders mix the culinary element with the tourism product to diversify the touristscape (Spilková & Fialová, 2013). In combination with the foodscape, a rural touristscape of this kind offers new features to tourists' experience (Robinson & Clifford, 2012; Wan & Chan, 2013). However, tourists' motivations concerning food are more complex.

Some scholars believe that in food tourism the primary motivating drive of a tourist is the food (Smith & Costello, 2009). Such consumers mostly (but not always) resemble post-materialistic tourists who focus on self-actualisation, self-fulfilment, and deep immersion into the experiences of tourism products (Blamey & Braithwaite, 1997; Wang, 2016) for the development of positive emotions and psychology. On the contrary, some suggest that food consumption could be secondary (Bertella, 2011; Presenza & Simone, 2012). Tourists for whom the local cuisine is a secondary motivational component take a shallow dip into the cultural values of their locale and often remain indifferent to the consequences of their activities either on themselves or the host communities (Porritt, 2012; Sharpley, 2009, 2012).

The next section of this chapter further discusses the characteristics of such tourists in detail. Positive tourism, bringing hope, positive psychology and emotions, happiness and well-being for both the hosts and guests, has been extensively in tourism academia in the recent past (Christou, 2021; Filep & Pearce, 2013; Filep et al., 2016). However, how the foodscape promotes positive tourism in a ruralscape has not been investigated through the lens of post-materialism. Therefore, this chapter attempts to unveil how the foodscape in rural areas acts as a positive tourism site to bring hope, happiness and other positive psychological components to the lives of post-materialistic tourists and their host communities. The chapter begins with a brief discussion of the significant characteristics of post-materialistic tourists in general, followed by the nuances of food tourism in rural cultural setups and its relation to the development of positive psychology for both hosts and guests (especially post-materialistic tourists). The chapter concludes with a practical experience from a case study on the Silk Route destination of East Sikkim in India, typically a rural touristscape with the trace of culinary tourism products blended with local culture.

Post-Materialistic Tourists

The term post-materialism was coined by Inglehart (Inglehart, 1971, 1977, 1990, 1997, 2008; Inglehart & Flanagan, 1987). His theory was primarily based on Maslow's (1954) hierarchy of needs. In his theory of human motivation, Maslow categorised human needs into five hierarchical classes: physiological needs, safety needs, belongingness and love needs, esteem needs, and self-actualisation. While the characteristics of post-materialistic individuals coincide with the highest order of human needs, the need for self-actualisation (Blamey & Braithwaite, 1997), materialist people are more occupied with lower order needs like physiological, safety and financial needs.

The general consumption pattern reflects a value shift from materialistic to post-materialistic individuals regarding ego-centred and other-centred tourism preferences. While explaining the two forms of travel and tourism opportunities through the lens of a pleasure-seeking mentality – the *summum bonum* – Fennell (2018: 385) argued that ego-centred tourism is preferred by individuals who are more into fulfilling personal leisure and pleasure requirements. On the other hand, in other-centred tourism, individuals seek to experience 'exotic' flavours in natural and cultural entities like physical landforms, cultural traditions, lifestyles, and so on. Moreover, tourists with post-materialistic values have a positive attitude towards conservation as they are more aware of the environment and alternative forms of tourism products (Chao & Chao, 2017).

Food Tourism in Cultural Ruralscapes

A significant debate exists within the growing research on food and tourism. Some scholars believe that food may be the primary motivation when tourists choose a destination (Hall *et al.*, 2004). Some also believe that food preference embraces just a partial section of the whole gamut of tourist motivation to visit a destination (Hjalager & Richards, 2003; Lau & Li, 2019). While the first proposition involves considering food tourism as a special interest tourism product, the latter suggests a holistic approach to exploring tourist motivation, which often comprises a multi-dimensional set of interests and expectations. Food tourism is usually found to develop as a factor or part of the broader context of tourism products, so researching food tourism in isolation or as a special interest tourism product is typically considered a 'myopic approach' (McKercher *et al.*, 2008: 138). In contrast to the definition of food tourism in terms of tourists' primary motivation or activity-based experiences, it is defined as any tourist experience related to culinary resources (Presenza & Del Chiappa, 2013).

With growing food tourism research, the discourse has gradually experienced a cultural turn (Everett, 2012). Food is increasingly becoming an intrinsic and inextricable component of culture tasted by tourists at a destination. Tourists experience the 'other' while consuming the culinary culture (Horng & Tsai, 2010). While trying the food either in an urbanscape or ruralscape, tourists ultimately taste the culture embedded in the food. However, what gives the foodscape its 'exotic' quality differs between the urbanscape and the ruralscape. The urbanscape is typically characterised by a relatively modern and affluent lifestyle, urban mentality (weaker community feelings), built-up environment, developed city morphology (streets, markets, luxury, etc.), technological interventions, restaurants, pubs, clubs, and so on (Lew & Cartier, 2004). In contrast, the ruralscape is characterised by vast agricultural or open land, remoteness, decreased noise, decreased pollution, culture, history, ethics, serenity, tranquillity, solitude, less-explored natural landscapes, traditional lifestyles, rural mentality (strong we-feelings), less technological interventions, and so on (Chaudhary & Lama, 2014; Saxena, 2012). Food tourism in rural areas gets commercialised in association with the main rural tourism product as a small-scale business, focusing more on regional development, promoting culture, and product diversification (Ellis *et al.*, 2018; Kim & Iwashita, 2016).

Motivation

The motivations of tourists visiting a culinary destination range from seeking authenticity, wanting to experience culture, having a need for status, luxury, self-esteem and self-actualisation, to participating in food or gastronomic events. So, the promotion and design of food events at a tourist destination invariably depend on visitors' motivations – from the physical to the psychological, and from physiological and security needs to social and cultural ones (Ellis *et al.*, 2018). However, in the process of embracing alternative hedonistic values, post-materialistic tourists are drawn to other-centred tourism products, inherent in the ruralscape as an opportunity to escape from homogenised and McDisneyised everyday life (Ritzer & Liska, 1997; Santini *et al.*, 2011).

The motivation of such tourists resembles Cohen's tourist type, who phenomenologically remain in the experiential or experimental mode and are internally dissatisfied with their societal 'centre'. They set out to experience or experiment with the 'centre' of a new, 'exotic' society and culture (Cohen, 1979). Their quest for otherness (MacCannell, 1973) takes them closer to the local food in the ruralscape, which often represents the place's authentic identity in terms of history, story, cultural traditions and symbols. Post-materialistic tourists interact with and consume such elements to create their experiences, motivated by engagement, participation, and their search for authenticity. Therefore, the foodscape of a rural destination directly widens the scope to affect tourists' positive psychology in terms of experiencing an 'exotic' culture and achieving internal happiness, self-actualisation and extreme pleasure.

Local cuisines and cultural value

The foodscape in a rural area can be considered a terroir product. A foodscape is argued to be constituted by the historical, cultural, and political landscape of a particular destination, blended with the food experience through official marketing (Amore & Roy, 2020). Furthermore, terroir products are conceived of as blending production processes with local resources in synergy with that destination's physical and cultural environment (Santini *et al.*, 2011). Hence, in the form of terroir products, regional cuisines appear as the 'cultural marker' of a specific territory (Santini *et al.*, 2011: 175) and provide ample scope to promote other-centred positive tourism.

Food brings in the context of a sense of place (Sims, 2009) and the territory's identity where the foodscape is promoted. As a symbolic language, the local cuisine, its preparation style and traditions surrounding its consumption reflect the origin and evolution of the Indigenous culture in many ways (Cianflone *et al.*, 2013; Montanari, 2009; Stringfellow *et al.*, 2013).

Authenticity

The interplay between the foodscape and tourists is typically characterised by cultural experiences, sensory appeal, interpersonal relations, excitement and health concerns (Kim & Eves, 2012). Sensory appeal refers to the physical involvement of tourists in the form of tasting, smelling and touching the food. Interpersonal relations enable the visitors to socially interact with the local people to experience their 'exotic' way of life. Health concerns centre around the well-being of the tourists while they consume the local food. However, the interplay is not necessarily similar across foodscapes. Sometimes tourists may also avoid local food due to health concerns, hygiene, and personal digestive tolerance levels. Tourists experience culture through authentic culinary practices and learn through sensing the place and exploring the history and stories embedded in the food praxis. However, authenticity gets altered over time based on the demands of tourists and demonstrations by hosts. This interplay generates hope among the hosts and guests through experiences mutually accepted to be authentic, which ultimately culminates in positive tourism. It is argued that the happiness and well-being of tourists come more from experiential purchases than materialistic consumption (Gilovich *et al.*, 2015).

Culinary product development and marketing strategy in the ruralscape

Promoting culinary heritage as a consumable product can directly benefit local communities and boost the economy (Bessière, 1998). In addition, it may contribute significantly to environmental and social well-being in regional development (Hall & Gössling, 2016). Therefore, valorising food as cultural heritage has become a new trend involving promoting culinary tourism in destinations where local stakeholders face the identity problems and challenges that result from destination competitiveness. In such settings, stakeholders strive to encourage the new form of gastronomy based on their culinary heritage to make it economically beneficial (Bessière, 1998).

It is evident in the ruralscape that tourism stakeholders imitate each other in the preliminary stage while designing the culinary component as a subset of the rural tourism product. With time and increasing profit, when more stakeholders start offering the same kind of food to the tourists, product differentiation becomes inevitable to survive market competition. In such a situation, competitors redesign the typical culinary products in the ruralscape to blend their cultural identities and uniqueness (Santini et al., 2011). However, overdiversification of food may confuse tourists and lead to subsequent dissatisfaction because of visitors' inability to choose the authentic product (Walsh, 1994).

The Sikkim Experiences

The following sections provide empirical evidence to support the theoretical context discussed earlier. The case study of Sikkim reflects how the food tourism of a ruralscape acts as an agent to bring happiness, positive psychology and well-being to the host and guest communities.

Methodology

Selection of study area

Sikkim, a tiny state located in the Eastern Himalayan region on the Silk Route, was selected as the study area. With an area of 7096 km^2, Sikkim offers tropical to temperate to alpine climatic regions mixed with the rich cultural landscape of three significant communities: the Lepcha, Bhutia and Nepalese. The primary source of revenue for this state is tourism. Among the several forms of tourism products, rural tourism is vital throughout Sikkim (Chaudhary & Lama, 2014).

The Silk Route of East Sikkim has recently emerged as a new offbeat tourist destination branded by the Sikkim government on its official tourism website. This new tourism circuit has been observed to have positive effects on the local people in the form of economic benefits and improved quality of life and happiness (Manhas et al., 2014). Tourists are also increasingly visiting this route in order to experience the 'exotic' nature and culture of the rural setup. This historical route once used to be the trading corridor between India and Tibet. The trading through this route was active before the Chinese invasion in Tibet became a geopolitical issue. The path once extended from Lhasa in Tibet to Kalimpong in the West Bengal state of India through the mountain pass, Jelep-la, when trading was taking place. Figure 6.1 shows the location of the Silk Route of Sikkim. The Silk Route has evolved into a tourism circuit comprising a series of destinations located at different altitudes in synergy with dispersed rural village clusters.

Sample selection

A total number of 100 homestays were surveyed along the Silk Route. The homestays are located in different villages, and Table 6.1 gives their distribution. The homestays were selected to cover almost all the units. No specific categorical criteria were decided upon to stratify the homestay

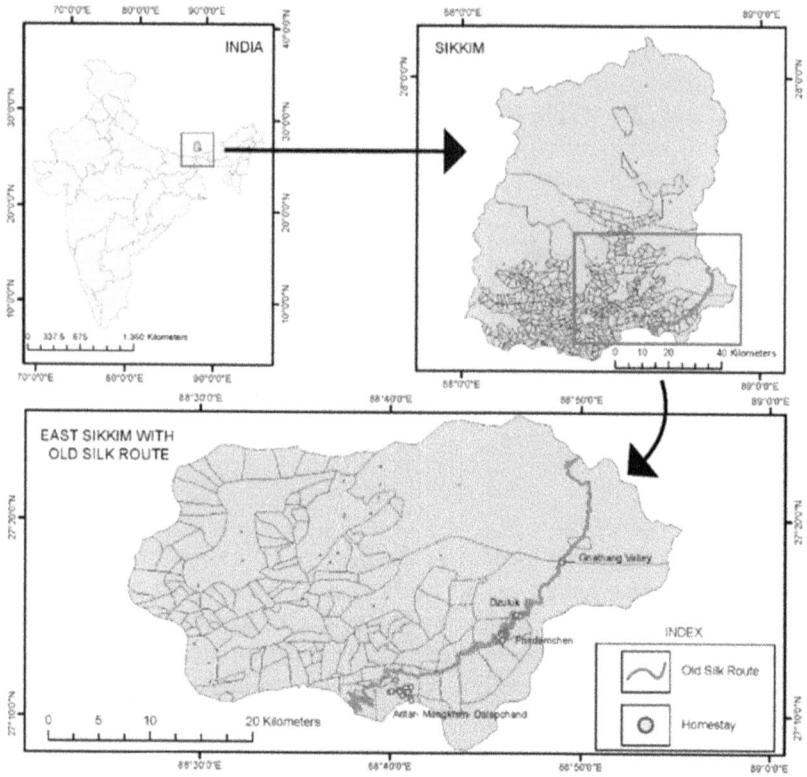

Figure 6.1 The Silk Route of Sikkim

Table 6.1 Village-wise distribution of homestay units and tourists surveyed

Villages/ village clusters	Total no. of homestay units	No. of leased-out homestays	Percentage of leased-out homestays (%)	No. of tourists at homestay units	No. of tourists at non-homestay units
Aritar–Mangkhim–Dalapchand	32	10	31.25	5	15
Phadamchen	44	28	63.64	10	40
Dzuluk	16	7	43.75	5	10
Gnathang Valley	8	3	37.50	5	10

units during the selection process. Therefore, every homestay had an equal chance of being selected, irrespective of their location, duration of operation (how long they have been active in the business), nearness to the viewpoints, room quality and leased or non-leased property status. At the initial stage, a list of homestays was prepared based on their names and the information available on different government and travel websites like Sikkim Tourism (www.sikkimtourism.gov.in/Public/index), Travelguru (www.travelguru.com) and Make My Trip (www.makemytrip.com). Based on references from the listed homestay owners, other homestays were traced out and surveyed during the fieldwork. The additional homestays were obtained from the clusters of non-homestay accommodation units using the snowball sampling technique. Because of the presence of some temporary homestays, the exact number of homestays excluded from the survey is uncertain. Based on the residents' references, almost 95% of homestays were surveyed. The table also shows the status of the homestay units in terms of being leased out or not. While selecting the accommodation units, only the homestays were sampled purposefully, and the other categories like lodges and resorts (n = 6) were excluded from the sample.

Besides the homestay units, 100 tourists who were staying at homestay units at different locations were interviewed. Most of the tourists were surveyed at Phadamchen based on availability. It was evident from the fieldwork that tourists generally prefer to stay at Phadamchen as it is situated at the midpoint of the Silk Route and has the maximum number of homestays compared to other locations along the Silk Route. Because the number of homestays is not uniform across the villages within this cluster, when surveying the homestay owners, many homestays were found to be vacant. Tourists over 30 were considered for the interviews. Tourists were found to be mostly visiting the Silk Route with family and friends. When asked about their age during interviews, it was revealed that at least 90% of them belonged to the over-30 age group. In the preliminary rounds of the tourist survey, some respondents below the age of 30 were also interviewed. It was found that either their answers were biased by their elderly companions, or they were reluctant to respond. Another reason was that most of the tourists below the age of 30 were teenagers and travel decisions had been made on their behalf by elderly family members. Of the sampled tourists, 40% were female and 60% were male. Tourists were interviewed at different hotel lawns, tea stalls and viewpoints. It is important to note that samples were selected from the homestay and non-homestay units for a comparative analysis of tourists' opinions regarding the importance of authentic culinary experiences.

Study approach and data collection

The case study adopts a qualitative approach in general and narrative methodology in particular. In-depth interviews were conducted with the 100 homestay owners. They were asked to express their opinions on the potential of their traditional food in association with the homestay operations. They were also asked to explain how they blended authentic culinary traditions with traditional accommodation experiences. In addition, they explained their reasons for leasing out the homestay units. Homestay owners were prompted to share their experiences of feedback from tourists regarding their satisfaction and happiness. Furthermore, how the homestay stakeholders benefitted from such tourism activities was also recorded.

Tourists were asked to share their primary motivation to visit the Silk Route, preferences when choosing accommodation units and local cuisines, and willingness to pay more for an authentic experience. Tourists were also asked how their experiences made them feel while consuming rural tourism products and food. Finally, they were asked about their sensitivity to the host community and their traditions to categorise them as materialist and post-materialistic tourists. The interviews were conducted in the Hindi language for homestay owners and Bengali for the tourists and were then transcribed verbatim into English. Related secondary data on tourists' arrivals were collected from the permit office under the police department at Rongli Bazar.

Tourism in the ruralscape of the Silk Route

This circuit destination is only open for Indian nationals with restricted travel permits because of border security issues with China. Domestic tourists must acquire travel permits from Rongli Bazar marketplace, the starting point of this tourism route. A tourist can obtain a travel permit for the Silk Route only through tour operators (primarily the locals). Because of the limited allowance for vehicle movements and tourists in this route, sometimes tourists have to wait one or two days for their turn. Thus, adjacent to the Rongli Bazar, a cluster of accommodating rural destinations and sightseeing locations have also emerged, popularly known as the Aritar–Mangkhim–Dalapchand village clusters. The route is characterised by a sharp altitudinal rise from 1530 m at the Rongli Bazar to 3705 m at Gnathang Valley and a fall in temperature and oxygen level within 60 km. Therefore, acclimatising tourists to the altitude, temperature, and oxygen level is a crucial issue on this route. This has culminated in the development of two intermediary resting destinations: Phadamchen (2050 m) and Dzuluk (2760 m).

Although the first evidence of tourists in this region cannot be traced, tourist arrivals have been appropriately recorded since 2007–2008 with the enforcement of travel permits. The collected tourist arrival data reflects that in 2017 (from July to December), tourist arrivals amounted to 15,685, which increased to almost 63,400 in.2018 (from January to September). The first homestay in the Silk Route was recorded as being developed in Dzuluk. Eventually homestays increased in all destinations along the route, especially after 2015 due to the sudden rise in tourist arrivals. The number reached more than 100; some of these are officially registered with the government of Sikkim and some are not. It was found that the accommodation units now frequently utilise the brand value of being a homestay. On asking a small group of homestay owners at Phadamchen about the effects of authentic and unauthentic homestay unit development, one of them stated:

> [W]hen we started the homestay business, we decided to maintain the sanctity of the term homestay by offering our culture and earning good life for us, and still we try hard to maintain that standard. Now, you can see dozens of hotels calling them[selves] homestays without knowing what to offer and why to offer. . . . Actually, they malign the brand name, which has done a lot in bringing money into their lives and [a] smile on their faces.

This reflects that the host community has received positive benefits like money and improved living standards from the homestay business. They also hope for further gains, although the unauthentic homestay owners may affect this adversely in the long run.

The foodscape in the touristscape

The Silk Route's primary attraction is nature-based, nestled as it is in the less-visited ruralscape, characterised by dispersed settlement clusters separated by intermittent stretches of dense vegetation cover on the mountains, which are flanked by several narrow but deep stream channels. Tourists are primarily found heading towards Gnathang Valley, the last destination on the circuit, offering a scenic view of a mountain-locked wide glacial valley at an altitude of more than 3000 m.

Intermediate destinations like Phadamchen and Dzuluk primarily attract tourists with their accommodation facilities and, secondarily, the scenic beauty of their physical and cultural landscapes. The accommodation units have evolved to take the form of homestays, offering food and beds within the houses of the local inhabitants. There is a growing trend of homestay owners leasing out their accommodation units to reduce the risk of economic loss due to unsold rooms, mainly during the off seasons. Located in extremely remote areas, the villagers of the Silk Route do not have much access to amenities like telecommunications and internet facilities. They have to depend on their physical connections with the tour operators and the tourism players of the Aritar–Mangkhim–Dalapchand village cluster to get their homestays booked. As a result, leaseholders from outside the area have appeared as a new type of player in this tourism market. However, not all homestay owners follow that path, and some prefer to run the homestay by themselves to reduce profit leakage to outsiders. Forty-eight homestays (almost 50%) were found to be leased-out properties during the field visit. According to operators:

> [The y]oungsters of our family are now doing government jobs. So, who is going to take care of the homestay business? Being aged, we find it convenient to lease out the property to outsiders. We get handsome money from them every year, and we don't have to take the headache of bringing the tourists.

However, remarkably, no homestay owners were found to be leasing out their kitchens, whether authentic or unauthentic. On asking some homestay owners about the reasons, they replied:

> [F]ood is the basic need of tourists. We can cook for them and get employment throughout the year. We can also earn extra and don't have to maintain two kitchens [laughing loudly].

They were observed to be more concerned about the economic benefit of tourism and the resulting improved quality of life.

Out of the remaining 52 homestays, 29 were not leased out, but the owners were indifferent to authentically promoting local cuisine. One such homestay owner opined:

> [W]e started the homestay quite long back and want to carry out the same by ourselves only. If we lease out the property, there will be no difference between a homestay and a hotel. We don't want a hotel business at all. However, with the market competition from the leased-out homestays, we provide rooms and food at a cheaper rate. . . . [V]ery few tourists like traditional food because we have to charge high for our efforts to prepare the special traditional menu which we don't consume every day.

Only 23 owners out of the 52 emphasised the authentic experiences for tourists and positive psychology in promoting culinary traditions themselves. According to one:

> [W]e want quality, not quantity. We charge high for the authentic experience, and we believe authentic experience comes at some cost. Therefore, we maintain the authenticity of food and room and welcome only those tourists willing to pay high for the exotic authentic tradition. We ensure that every tourist will return happily with good memories, extreme satisfaction, and future hope to revisit. As we charge high, we receive a lesser number of tourists in comparison to other homestays, but still, we earn sufficient to maintain a good life and get motivated to serve better in future.

Therefore, such homestays are typical examples of tourism stakeholders who believe in developing the positive psychology of tourists by adding value to the local culture in culinary form. In other cases, the food is losing its terroir authenticity because of the continuous modification and alteration of the local culinary menu. This is caused by the steady demand of tourists who prefer not to engage with the 'exotic' gastronomic flavour for a prolonged time, even at an unfamiliar destination.

Visitor patterns and market potential

Food in the Silk Route tourism circuit works both as a primary and a secondary or incidental motivating factor for tourists. Of those surveyed, around 70% were categorised as materialistic due to their prioritisation of and concerns about expenses. They often travel in groups to reduce per-head expenditure. In a group interview with them, they stated:

> [W]e love nature. And [we] expend much on reaching the sights, because the car rental is very high. So, we try to cover the sightseeing locations as much as possible within less time. . . . [W]e spend less on accommodation, as we travel mainly in the group. So, spending much on accommodation is unproductive as we can spend that money to cover more destinations. . . . [W]e are minimalistic in food and prefer the regular menu we consume at our places. However, we don't mind if we are served a unique local dish, but not all the time. Because, you know, they charge high for that, and we love our regular items.

From their comments, it is evident that they are less interested in authentic cultural experiences and are often less immersed in the cultural heritage. In the present example, the happiness, joy, satisfaction and other positive psychological emotions of such materialistic-minded tourists are influenced by nature-based tourism and less expensive tourist facilities.

However, the contribution of such tourists in valuing local culture and encouraging locals to showcase the authentic culinary tradition is almost insignificant. A group of 10 friends accompanied by their family members were asked about their opinion on how they look at authentic local dishes. They responded:

> [W]hat do you mean by authenticity? Is it about their handicrafts? We didn't find any such product here. . . . Ok, now we understand. See, we can get it anywhere, you know? Google will help [laughing]. We are not interested in local food at extra cost. We focus on sightseeing mainly. And you know, nowadays, nothing is left authentic because of business. Instead, locals should learn how to cook modern cuisines as per the demand of tourists . . .

The above response clearly shows their indifference to authentic food. Prompted about authentic food instead of handicrafts, they understood this was about the local cuisine. However, their comments exhibit their limited interest in the authenticity of the local food.

In contrast, the remaining 30% of tourists wanted to experience everything more authentically. Their happiness was more associated with immersion into the cultural landscape, in which the foodscape is a significant component. One family staying at an authentic homestay said:

> [O]ur preference is solitude. We get bored of the daily urban crowd. So, when we need a break, we select a rural area away from city life. It is so peaceful and refreshing, and, significantly, people are so lovely and welcoming. . . . [E]very time we get amazed by their culture of simplicity and hospitality; we learn to be happy with minimum material needs.

Another couple who had a hectic and affluent lifestyle opined:

> [W]e just don't like the crowd. We love to try new experiences from the Indigenous people, whether food or lifestyle. People nowadays don't understand the intrinsic reward value of experiencing simplicity and traditions. We know, staying at an authentic homestay is expensive. But they should consider it a token of appreciation for what they learn from the traditional values. We taste their food but hardly make justice to their efforts in arranging that by giving a price. Those are priceless.

Interestingly, when one of the authentic homestay owners was asked about the nature of their relationship with the guests, one lady replied:

> [T]hey are amazing. I can't separate tourists from the other family members most of the time. They come to my kitchen, learn how it is cooked, and we do lots of fun together. Still, I have regular contact with them. Sometimes they say that they serve our traditional food to their guests as well [smiling with bright eyes]. We, the family members, and the guests find such interactions very relaxing and beneficial, both economically and socially.

It is evident from these narratives that such tourists are heavily into the intrinsic value of the ruralscape and its traditional food. Their willingness to experience culture and value symbolic representations like food make them post-materialistic tourists. The narratives also reflect how the host community receives incentives from such post-materialistic tourists to promote the local culture and food and boost their economy and quality of life.

Discussion

The empirical evidence presented here has a solid contextual relationship with the theories discussed in the first half of this chapter. It has been shown that the host community generates a significant amount of revenue from the homestay business irrespective of their engagement in promoting authentic food experiences for the tourists. This connects with theoretical arguments that improved quality of life and overall well-being of locals are significant effects of food tourism, leading to happiness in the host community (Matteucci & Filep, 2017).

It is evident that some stakeholders are keen to maintain the authenticity of the homestays and local cuisines, while some are not. This aligns with the argument put forth by MacCannell (1973), who rightly alleged that hosts were responsible for serving a modified experience to the guests. On the other hand, the pressure homestay owners to compromise authenticity (due to the demands of tourists for lower accommodation and food prices) support the viewpoints of Boorstin (1961), who asserts that tourists prefer pseudo-events. These findings chime with the position that authenticity results from the interplay between the foodscape and tourists (Kim & Eves, 2012).

The Sikkim case study reveals that authentic homestay owners emphasise making more space for tourists to experience the local lifestyle, food preparation traditions, and customs significance. Such findings dovetail with scholars' arguments, suggesting an undeniable and inextricable link between food and culture, which differs from other forms of conventional tourism products (Silkes *et al.*, 2013; Updhyay & Sharma, 2014). The blending of culture and food makes it the 'cultural marker' of a specific territory (Santini *et al.*, 2011: 175). Furthermore, the promotion of cultural authenticity by such homestay owners supports the argument that local cuisines are often seen as a cultural relic reflecting the cultural heritage of a tourist destination (Metro-Roland, 2013). In this regard, the 'geography of taste' inevitably represents a place with its foodscape in terms of the relationship between local cuisine and cultural values (Kim & Ellis, 2015; Marcoz *et al.*, 2016).

It is also apparent from the narratives that some tourists prefer the authentic experience of homestays, including local cuisines, while others do not. Those indifferent to the authentic foodscape coincide with Fennell's (2018) ego-centred tourism preferences. In another way, their experiences fall under the categories of entertainment and education, as argued by Santini *et al.* (2011). In entertainment experiences, tourists participate passively in the reality of the tourism experience rather than being immersed in it. On the other hand, educational experiences involve them actively in the reality in front of them. Tourists who are less attentive to cultural values may be called materialists with reference to the work of Inglehart (1977) and Maslow (1954), who argued that such people are occupied with lower order needs like physiological, safety and financial needs. Their experiences of happiness are short-term and of the hedonic type (Christou, 2021).

For those who commented in favour of an authentic foodscape in rural areas, their experiences have escapist and aesthetic qualities. Escapists actively immerse themselves in their surroundings, and they expect cultural authenticity from the tourism product in its purest form, as found in the empirical evidence. The aesthetic experience comes from tourists' active immersion into reality. Soper (2008: 571) argued that such tourists consider 'alternative hedonism' as a sustainable option as opposed to the affluent and consumerist approach of most tourists' consumption. Such tourists might be called post-materialists, preferring higher-order needs like self-actualisation, as Maslow (1954) proposed. Post-materialists mostly opt for other-centred tourism products as a form of alternative hedonism to experience the pleasure of extreme goods – the *summum bonum* (Fennell, 2018). Valuing authentic culture in a rural foodscape through the lens of post-materialism becomes an alternative hedonism for such tourists. Their experiences of happiness are argued to be more of the eudaimonic type, which lasts for a longer time (Christou, 2021).

Nevertheless, the above discussion argues that food tourism in a ruralscape may act as positive tourism, bringing hope, happiness and positive psychology to both the host and the guests. Filep *et al.* (2016) explained that the concept of positive tourism is grounded on two fundamental theories, humanistic psychology and positive psychology. While humanistic psychology embraces the 'good life, individual growth and achievements, authenticity and personal responsibility' as its central theme, positive psychology is about the processes and mechanisms of an individual's flourishing and self-actualisation, and what makes life worthwhile (Filep *et al.*, 2016: 5). They also put forward that positive tourism affects positive psychology at three levels: first, the tourists' experience; second, at the host level; and third, at the level of tourism workers. In brief, positive tourism brings positive psychological effects at different levels for its stakeholders in terms of the good life, self-actualisation, flourishing, happiness and a meaningful life.

The case study revealed that although most (though not all) of the materialistic tourists experience the cultural foodscape in an incidental form, their overall experience at a ruralscape often makes them happy and engenders positive psychology with the natural beauty of the rural landscape. Similarly, hosts gain hope of economic benefit and the good life and happiness as a result. Moreover, through food tourism and its embedded culture at a rural destination, post-materialistic tourists get scope for self-flourishment and self-actualisation.

Conclusion

The present chapter has strived to demonstrate how food tourism may act as a source of happiness, benefits, and hope for both hosts and guests. It has also focused on the derivation of positive psychology among tourists while experiencing the foodscape in a ruralscape. The chapter reveals that post-materialistic tourists value the Indigenous culture of their locale and its symbolic demonstration. On the other hand, materialistic tourists contribute primarily to the economic gain of the host community.

Post-materialistic tourists and their desire for and impact on the foodscape deserve further attention from researchers. Furthermore, the motivating factors for post-materialistic tourists, their consumption behaviour and positive psychology are broader themes that need further exploration in future tourism research. This chapter appears like the tip of the iceberg, unveiling more dimensions of post-materialistic tourism that are yet to be explored in the long run.

Acknowledgements

We extend our sincere gratitude and thanks to all the respondents who cordially cooperated with us for the factual mining process. Without their help, this chapter's significant discussion section would have remained incomplete. Moreover, the first author is also thankful to the Presidency University for sanctioning the research grant to fund the extensive fieldwork along the Silk Route of East Sikkim.

References

Amore, A. and Roy, H. (2020) Blending foodscapes and urban touristscapes: International tourism and city marketing in Indian cities. *International Journal of Tourism Cities* 6 (3), 639–655.

Bertella, G. (2011) Knowledge in food tourism: The case of Lofoten and Maremma Toscana. *Current Issues in Tourism* 14 (4), 355–371.

Bessière, J. (1998) Local development and heritage: Traditional food and cuisine as tourist attractions in rural areas. *Sociologia Ruralis* 38 (1), 21–34.

Blamey, R.K. and Braithwaite, V.A. (1997) A social values segmentation of the potential ecotourism market. *Journal of Sustainable Tourism* 5 (1), 29–45.

Boorstin, D. (1961) *The Image: A Guide to Pseudo-Events in America*. New York: Atheneum

Chao, Y.-L. and Chao, S.-Y. (2017) Resident and visitor perceptions of island tourism: Green sea turtle ecotourism in Penghu Archipelago, Taiwan. *Island Studies Journal* 12 (2), 213–228.

Chaudhary, M. and Lama, R. (2014) Community based tourism development in Sikkim of India—A study of Darap and Pastanga villages. *Transnational Corporations Review* 6 (3), 228–237.

Christou, P.A. (2021) *Philosophies of Hospitality and Tourism: Giving and Receiving*. Bristol: Channel View Publications.

Cianflone, E., Di Bella, G. and Dugo, G. (2013) Preliminary insights on British travellers' accounts of Sicilian oranges. *Tourismos: An International Multidisciplinary. Journal of Tourism* 8 (2), 341–347.

Cohen, E. (1979) A phenomenology of tourist experiences. *Sociology* 13 (2), 179–201.

Ellis, A., Park, E., Kim, S. and Yeoman, I. (2018) What is food tourism? *Tourism Management* 68, 250–263.

Everett, S. (2012) Production places or consumption spaces? The place-making agency of food tourism in Ireland and Scotland. *Tourism Geographies* 14 (4), 535–554.

Fennell, D.A. (2018) On tourism, pleasure and the *summum bonum*. *Journal of Ecotourism* 17 (4), 383–400.

Filep, S. and Pearce, P. (eds) (2013) *Tourist Experience and Fulfilment: Insights from Positive Psychology*. Abingdon: Routledge.

Filep, S., Laing, J. and Csikszentmihalyi, M. (eds) (2016) *Positive Tourism*. Abingdon: Routledge.

Fredrickson, B.L. (2001) The role of positive emotions in positive psychology: The broaden-and-build theory of positive emotions. *American Psychologist* 56 (3), 218–226.

Gilovich, T., Kumar, A. and Jampol, L. (2015) A wonderful life: Experiential consumption and the pursuit of happiness. *Journal of Consumer Psychology* 25 (1), 152–165.

Gump, B.B. and Matthews, K.A. (2000) Are vacations good for your health? The 9-year mortality experience after the multiple risk factor intervention trial. *Psychosomatic Medicine* 62 (5), 608–612.

Hall, C.M. and Gössling, S. (2016) *Food Tourism and Regional Development: Networks, Products and Trajectories*. Abingdon: Routledge.

Hall, C.M., Sharples, L., Mitchell, R., Macionis, N. and Cambourne, B. (eds) (2004) *Food Tourism around the World*. Abingdon: Routledge.

Hillman, P., Moyle, B.D., Weiler, B. and Che, D. (2016) The impact of tourism on the quality of life of local industry employees in Ubud, Bali. In S. Filep, J. Laing and M. Csikszentmihalyi (eds) *Positive Tourism* (pp. 148–164). Abingdon: Routledge.

Hjalager, A.-M. and Richards, G. (2003) Still undigested: Research issues in tourism and gastronomy. In A.-M. Hjalager and G. Richards (eds) *Tourism and Gastronomy* (pp. 224–234). London: Routledge.

Horng, J.-S. and Tsai, C.-T. S. (2010) Government websites for promoting East Asian culinary tourism: A cross-national analysis. *Tourism Management* 31 (1), 74–85.

Inglehart, R. (1971) The silent revolution in Europe: Intergenerational change in post-industrial societies. *American Political Science Review* 65 (4), 991–1017.

Inglehart, R. (1977) *The Silent Revolution: Changing Values and Political Styles among Western Publics*. Princeton: Princeton University Press.

Inglehart, R. (1990) *Culture Shift in Advanced Industrial Society*. Princeton: Princeton University Press.

Inglehart, R. (1997) *Modernization and Postmodernization in 43 Societies*. Princeton: Princeton University Press.

Inglehart, R.F. (2008) Changing values among Western publics from 1970 to 2006. *West European Politics* 31 (1–2), 130–146.

Inglehart, R. and Flanagan, S.C. (1987) Value change in industrial societies. *American Political Science Review* 81 (4), 1289–1319.

Kim, S. and Ellis, A. (2015) Noodle production and consumption: From agriculture to food tourism in Japan. *Tourism Geographies* 17 (1), 151–167.

Kim, S. and Iwashita, C. (2016) Cooking identity and food tourism: The case of Japanese udon noodles. *Tourism Recreation Research* 41 (1), 89–100.

Kim, Y.G. and Eves, A. (2012) Construction and validation of a scale to measure tourist motivation to consume local food. *Tourism Management* 33 (6), 1458–1467.

Lau, C. and Li, Y. (2019) Analyzing the effects of an urban food festival: A place theory approach. *Annals of Tourism Research* 74, 43–55.

Lew, A.A. and Cartier, C. (2004) *Seductions of Place: Geographical Perspectives on Globalization and Touristed Landscapes*. Abingdon: Routledge.

MacCannell, D. (1973) Staged authenticity: Arrangements of social space in tourist settings. *American Journal of Sociology* 79 (3), 589–603.

Manhas, P.S., Kour, P. and Bhagata, A. (2014) Silk Route in the light of circuit tourism: An avenue of tourism internationalization. *Procedia - Social Behavioral Sciences* 144, 143–150.

Marcoz, E.M., Melewar, T. and Dennis, C. (2016) The value of region of origin, producer and protected designation of origin label for visitors and locals: The case of Fontina cheese in Italy. *International Journal of Tourism Research* 18 (3), 236–250.

Maslow, A.H. (1954) *Motivation and Personality*. New York: Harper & Row.

Matteucci, X. (2016) Tourists' accounts of learning and positive emotions through sensory experiences. In S. Filep, J. Laing and M. Csikszentmihalyi (eds) *Positive Tourism* (pp. 54–67). Abingdon: Routledge.

Matteucci, X. and Filep, S. (2017) Eudaimonic tourist experiences: The case of flamenco. *Leisure Studies* 36 (1), 39–52.

McKercher, B., Okumus, F. and Okumus, B. (2008) Food tourism as a viable market segment: It's all how you cook the numbers! *Journal of Travel & Tourism Marketing* 25 (2), 137–148.

Metro-Roland, M.M. (2013) Goulash nationalism: The culinary identity of a nation. *Journal of Heritage Tourism* 8 (2–3), 172–181.

Montanari, A. (2009) Geography of taste and local development in Abruzzo (Italy): Project to establish a training and research centre for the promotion of enogastronomic culture and tourism. *Journal of Heritage Tourism* 4 (2), 91–103.

Porritt, J. (2012) *Capitalism as if the World Matters*. Abingdon: Routledge.

Presenza, A. and Del Chiappa, G. (2013) Entrepreneurial strategies in leveraging food as a tourist resource: A cross-regional analysis in Italy. *Journal of Heritage Tourism* 8 (2–3), 182–192.

Presenza, A. and Simone, I. (2012) High cuisine restaurants: Empirical evidences from a research in Italy. *European Journal of Tourism, Hospitality and Recreation* 3 (3), 69–85.

Ritzer, G. and Liska, A. (1997) 'McDisneyization' and 'post-tourism': Complementary perspectives on contemporary tourism. In C. Rojek and J. Urry (eds) *Touring Cultures: Transformations of Travel and Theory* (pp. 96–112). London: Routledge.

Robinson, R.N. and Clifford, C. (2012) Authenticity and festival foodservice experiences. *Annals of Tourism Research* 39 (2), 571–600.

Santini, C., Cavicchi, A. and Canavari, M. (2011) The Risk™ strategic game of rural tourism: How sensory analysis can help in achieving a sustainable competitive advantage. In K.L. Sidali, A. Spiller and B. Schulze (eds) *Food, Agri-Culture and Tourism: Linking Local Gastronomy and Rural Tourism: Interdisciplinary Perspectives* (pp. 161–179). Heidelberg: Springer.

Saxena, G. (2012) Geographies of rural tourism: Current progress and paradoxes. In J. Wilson (ed.) *The Routledge Handbook of Tourism Geographies* (pp. 238–244). Oxon: Routledge.

Sharpley, R. (2009) *Tourism Development and the Environment: Beyond Sustainability?* London: Earthscan.

Sharpley, R. (2012) Does consumerism necessarily promote bad tourism? In T.V. Singh (ed.) *Critical Debates in Tourism* (pp. 54–61). Bristol: Channel View Publications.

Silkes, C.A., Cai, L.A. and Lehto, X.Y. (2013) Marketing to the culinary tourist. *Journal of Travel Tourism Marketing* 30 (4), 335–349.

Sims, R. (2009) Food, place and authenticity: Local food and the sustainable tourism experience. *Journal of Sustainable Tourism* 17 (3), 321–336.

Sirgy, M.J. (2002) *The Psychology of Quality of Life*. Dordrecht: Kluwer Academic.

Smith, S. and Costello, C. (2009) Culinary tourism: Satisfaction with a culinary event utilizing importance-performance grid analysis. *Journal of Vacation Marketing* 15 (2), 99–110.

Soper, K. (2008) Alternative hedonism, cultural theory and the role of aesthetic revisioning. *Cultural Studies* 22 (5), 567–587.

Spilková, J. and Fialová, D. (2013) Culinary tourism packages and regional brands in Czechia. *Tourism Geographies* 15 (2), 177–197.

Stringfellow, L., MacLaren, A., Maclean, M. and O'Gorman, K. (2013) Conceptualizing taste: Food, culture and celebrities. *Tourism Management* 37, 77–85.

Updhyay, Y. and Sharma, D. (2014) Culinary preferences of foreign tourists in India. *Journal of Vacation Marketing* 20 (1), 29–39.

Walsh, K. (1994) Marketing and public sector management. *European Journal of Marketing* 28 (3), 63–71.

Wan, Y.K.P. and Chan, S.H.J. (2013) Factors that affect the levels of tourists' satisfaction and loyalty towards food festivals: A case study of Macau. *International Journal of Tourism Research* 15 (3), 226–240.

Wang, Y. (2016) Social stratification, materialism, post-materialism and consumption values: An empirical study of a Chinese sample. *Asia Pacific Journal of Marketing and Logistics* 28 (4), 580–593.

7 Family Travel, Positive Psychology and Well-Being

Mona Mirehie and Iryna Sharayevska

Research has shown that travel as a form of family leisure results in various benefits for families. Families use travel to improve their relationships, escape from routine, create memories, improve communication and continue family traditions (Durko & Petrick, 2016; Shaw & Dawson, 2001; Zabriskie & McCormick, 2003). Over the last couple of decades, positive psychology and well-being have received increasing attention in assessing the benefits of travel. Studies have documented hedonic and eudaimonic elements of tourism experiences that contribute to an overall sense of well-being (e.g. Moal-Ulvoas, 2017; Neal *et al.*, 2007). Aiming at bringing together these two bodies of literature (i.e. family leisure and tourism well-being), in this chapter we present the findings of a qualitative study that explored travel and family well-being from a positive psychology perspective.

Family Travel

Travel experiences have been shown to be beneficial to individuals and families in many different ways. Travel allows individuals to escape their everyday routines, lower work-related stress, experience new things, practice control and freedom, improve their mental and physical health, and, as a result, increase their well-being (Chen *et al.*, 2013; Fritz & Sonnentag, 2006; Sonnentag & Fritz, 2007). Moreover, the benefits of travel are experienced not only at the individual level but by the entire family. Families reported increased bonds, connectedness and positive feelings; improved cohesion, relationships and loyalty between family members; and the creation of shared lifelong memories (Durko & Petrick, 2016; Gilbert & Abdullah, 2004; Kozak, 2010; Kozak & Duman, 2012; Shaw *et al.*, 2008; West & Merriam, 2009; Yun & Lehto, 2009). For instance, the participants in Lehto *et al.*'s (2009) study reported that holidays were viewed as quality time with the family that allowed for improvement of family communication and cohesion. Similarly, Lehto *et al.* (2012) found that family trips provided quality family time and opportunities to re-establish emotional bonds. Moreover, couples reported increased intimacy frequency, which was eight times higher when they were on holiday than when they were at home (Durko & Petrick, 2016). Experiencing such positive effects, families felt more satisfied with their relationships and families and were less likely to get a divorce or be separated (Durko & Petrick, 2016; Hill, 2000; Presser, 2000; Yun & Lehto, 2009). Their quality of life was also enhanced (de Bloom *et al.*, 2010; Dolnicar *et al.*, 2012; Fritz & Sonnentag, 2006; Strauss-Blasche *et al.*, 2000).

Furthermore, it has been found that children experience positive effects from travel too, such as enhanced development and socialisation, the acquisition of new skills (particularly sharing and getting along with others), fewer stress-related illness and behavioural issues, improved contentment with school, self and leisure life, and enhanced confidence levels and global life satisfaction (Gao *et al.*, 2020; Shaw & Dawson, 2001; West & Merriam, 2009). Children also reported feeling calm and relaxed, happier and recharged, as well as experiencing more opportunities to interact with peers and practice autonomy and independence (Mikkelsen & Stilling Blichfeldt, 2015), all of which resulted in enhanced overall wellbeing for the family.

Research on travel and extended family is more limited but also suggests positive outcomes from travel. Travel is often used to bring the members of extended families together to reconnect and strengthen bonds. For example, Kennedy-Eden and Gretzel (2016) studied the meanings of holidays for families living far away from each other. The study revealed that annual holidays with extended family have become a tradition that helps families to strengthen their bond and enhance family capital away from everyday routine, which the authors called 'system maintenance' (Kennedy-Eden & Gretzel, 2016: 14). Similarly, a study by Kluin and Lehto (2012) revealed that family reunion tourists manage complicated group dynamics and prioritise emotional factors over rational factors when they make decisions about travel because they put group interests before their personal preferences.

While family travels offer a volume of great outcomes, it is important to highlight that these experiences can also cause stress (Kennedy-Eden & Gretzel, 2016). For example, Rosenblatt and Russell (1975) discussed disruption of specific domestic routines, such as division of labour and space, while on holiday, leading to stress and conflict between family members. In addition, factors associated with travel, including illness, traffic and car issues, may serve as additional reasons for frustration and interpersonal conflict (Rosenblatt & Russell, 1975). Interestingly, some of these stressors may be mitigated by other factors. For example, a study by Smith *et al.*

(2017) suggested that couples with less travel experience were more likely to face conflicts in decision-making while those with more experience appeared to be able to avoid conflict. Although existing studies provide insight into the benefits and drawbacks of travel for families, the knowledge seems to be scattered; a holistic view of the positive and negative outcomes of family travel and an assessment of which are more significant – that is, the focal point of positive psychology – is missing. Hence, to address this gap in the literature, in this study, we employ positive psychology to explore family travel and its relationship with well-being.

Theoretical Framework

Positive psychology

Positive psychology conceptualises well-being as a combination of hedonic and eudaimonic components (Seligman, 2004, 2011). Accordingly, well-being is not solely the absence of mental illness. Rather it is the presence of some positivity, which consists of both subjective (i.e. mental health) and objective (i.e. quality of life) components that together help individuals flourish and achieve an enjoyable and worthwhile life (Seligman & Csikszentmihalyi, 2000). Seligman (2011: 16) proposed that well-being, as the centrepiece of positive psychology, consists of five facets: 'positive emotions, engagement, positive relationships, meaning, and accomplishment.... A handy mnemonic is PERMA.' Engagement refers to a positive state of mind that is experienced through full immersion in an activity (Diener & Seligman, 2002). Meaning and accomplishment refer to perceptions of leading a purposeful life and progressing towards personal goals, respectively (Seligman, 2011). Positive emotions (e.g. satisfaction, fulfilment) and positive relationships are also deemed to impact human well-being (Seligman, 2002).

Seligman (2004) clarified that positive emotions consist of both pleasure (conscious feelings) and gratification (elusive feelings). In alignment with in-depth assessments of affect and cognition in different fields (e.g. philosophy, psychology), Fennell (2009) highlighted the prominence of the concept of pleasure to tourism. Similar to Seligman, Fennell identified different types of pleasure such as emotional (e.g. satisfaction) and sensory (e.g. the unique scent of a destination) and discussed how different types of tourists (e.g. sex tourists versus volunteer tourists) experience different types of pleasure derived from distinct values. Furthermore, Fennell explained how experiences of pleasure are associated with three temporal phases of a travel experience: anticipation, on-site and recollection. Later, Fennell (2018) argued that pleasure is the main motive for travel and an ultimate goal in many people's leisure or even life experiences.

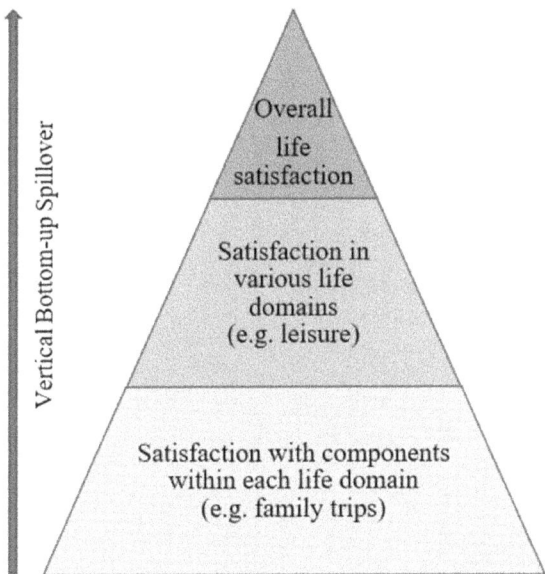

Figure 7.1 Hierarchical model of life satisfaction adapted to family travel
Source: Neal et al., 1999.

A majority of the studies within the realm of tourism and well-being have been founded on the bottom-up spillover theory of well-being (Figure 7.1), which postulates that overall well-being is the accumulation of satisfaction with major domains of life such as leisure, work and family life (Diener et al., 1985; Neal et al., 2007). Accordingly, the sense of well-being experienced from different elements (in this case family trips) of a specific domain (herein family leisure) spills over vertically to determine the overall satisfaction with that domain, which subsequently spills over to the topmost domain (overall life experience) and results in higher global life satisfaction (Neal et al., 2007).

Although many of the aforementioned studies on family travel do not claim any positive psychological theory as their conceptual framework, the eudaimonic and hedonic aspects of well-being are evident in the findings. Hence, with the aim of aggregating and synthesising the existing knowledge and solidifying the foundation for future work in this space, here we first outline the theories and measurement items used to investigate well-being in tourism contexts (Table 7.1); second, positioning ourselves within the positive psychological school of thought, we present the results of a qualitative study on travel and family well-being founded on the bottom-up spillover theory. To gain a well-rounded understanding of this phenomenon, eudaimonic and hedonic aspects of well-being found in previous tourism research (Table 7.1) have been used to guide the analysis.

Table 7.1 Theories and scales of positive psychology and well-being used in tourism research

Source	Instruments developed or adopted and adapted to tourism	Eudaimonic items	Hedonic items
Andrews & Withey (1976)	Global life satisfaction		Positive emotions
Kozma & Stones (1980, 1983)	Memorial University of Newfoundland Scale of Happiness (MUNSH)	Autonomy/control • Over what to do, how to do it, when to do it	• Experiencing happiness, joy, emotional involvement with people, fun, arousal/excitement, pleasure, euphoria
Kammann & Flett (1983)	Affectometer 2	Meaningful/purposeful life	• Feeling positive about oneself
Diener (1984)	Subjective well-being (SWB)	Personal growth	Positive relationships
Diener et al. (1985)	Satisfaction with Life Scale (SWLS)		• Having more quality/enjoyable than stressful social time
Watson et al. (1988)	Positive and Negative Affect Scale (PANAS)	Detachment	• Feeling close to family, friends
WHOQOL Group (1995)	Quality of life (QoL)	• Having physical and mental detachment from work demands	• Ability to re-establish dwindling relationships
Ryff & Keyes (1995)	Psychological well-being (PWB)	Mastery	• Ability to establish new relationships
Neal et al. (1999)	Satisfaction with leisure trips and satisfaction with life	• Having opportunities to broaden one's horizons • Having opportunities to seek physical and intellectual challenges	
Lyubomirsky & Lepper (1999)	Subjective Happiness Scale (SHS)	• Learning new things	Overall life satisfaction
Fredrickson (2001)	Broaden and build	Positive functioning (resilience and self-esteem)	• Feeling good about life despite ups and downs
Peterson et al. (2005)	Pleasure	• Self-acceptance, optimism, ability to bounce back from adverse life events	Satisfaction with different domains of life
Madrigal (2006)	Flow	Opportunities to be spontaneous	• Health
Sonnentag & Fritz (2007)	Recovery-Experience Scale	Engagement	• Income
Institute for Social and Economic Research (2009)	British Household Panel Survey (BHPS)	• Experiencing flow	• Accommodation
New Economics Foundation (NEF) (2009)	Emotional well-being (positive and negative affect)	Achievement Negative emotions	• Family • Employment status
	Eudaimonic well-being (psychological resources)	• Sadness, depression, stress	• Social life
Oliver (2010)	Satisfaction	Relaxation/rejuvenation	• Amount of leisure time Spending leisure time
Newman et al. (2014)	Leisure SWB – DRAMMA	• Taking time for leisure	Spending leisure time
Seligman (2011)	PERMA (positive emotions, engagement, positive relationships, meaning, and accomplishment)	• Using the time to relax • Doing relaxing things	

Note: There is a lot of overlap in items within different scales. To avoid redundancy, each item was only presented once in the table.

Methods

Data collection

In-depth semi-structured interviews were conducted with 18 adults who travel as a family on a regular basis. Convenience sampling was used to recruit the interviewees in the researchers' states of residence, Indiana and South Carolina, USA. An email was sent to potential participants to invite them to participate in the study. At the completion of each interview a gift card was granted to the interviewee. The interviews were conducted during spring 2020 (prior to COVID-19-related travel restrictions). Interviews were conducted in-person, via phone, or on Zoom and at a time that was most convenient for participants. The participants consisted of 18 full nesters (i.e. parents whose growing children still live with them), of which 13 were female and 5 were male (aged between 32 and 52 years old). The participants were highly educated, middle-class individuals, primarily white, and had between one and four children. Two were divorced, two were single, and 14 were married; one of the married participants was in a same-sex marriage and the rest were heterosexual parents (Table 7.2).

Participants were asked to reflect on their family travel experiences and the impacts of those experiences on their individual and collective well-being. To adhere to the overall theme of the book, in this chapter we solely focused on the positive side of the family trips related to well-being. Hence, Seligman's (2011) conceptualisation of well-being was used as a guiding theory for developing the interview questions. Examples of interview questions include, how does travelling with your family impact your life? What emotions do you have when you are travelling with your family? What does it mean for you to travel with your family? How do family trips impact your family relationships? Further probing was used when it was necessary to delve deeper into interviewees' narratives. The average interview length was one hour. Data collection ceased when concepts appeared to be saturated (i.e. no new relevant information was imminent) (Charmaz, 2006). Interviews were recorded and transcribed for analysis.

Data analysis

Thematic content analysis was used to analyse the data (Braun & Clarke, 2012). Eudaimonic and hedonic elements of well-being that were identified in the literature review were used as a guide for coding. The coding process started with manual incident-by-incident open coding. Once the researchers had discussed the codes and agreed upon the accuracy of the codes as well as consistency of the coding across the interviews, the most frequently repeated codes were used to guide the advanced phase of the analysis and to frame the findings. Once the coding was finalised, the most important codes were categorised into themes and sub-themes that explained families' well-being in relation to their travel experiences.

Table 7.2 Participants' profiles

Participant	Age	Marital status	No. of children	Race	Gender
1	47	Married	2	Caucasian	Female
2	44	Married	2	Caucasian	Male
3	41	Single	1	Caucasian	Female
4	43	Married	2	Caucasian	Male
5	37	Married	2	Caucasian	Female
6	38	Married	2	Caucasian	Female
7	43	Married	2	Caucasian	Male
8	32	Married	2	Caucasian	Female
9	42	Married	4	Caucasian	Female
10	52	Married	2	Caucasian	Male
11	43	Divorced	2	Caucasian	Female
12	49	Single	3	Caucasian	Female
13	48	Married	2	Caucasian	Female
14	48	Married	2	Caucasian	Male
15	46	Married	3	Caucasian	Female
16	47	Divorced	2	African American	Female
17	45	Married	1	Caucasian	Female
18	49	Married	1	Caucasian	Female

Results

When interviewees were asked why they took family trips, the most commonly mentioned reasons ($n = 17$) were visiting extended family to celebrate holidays and family events such as weddings, birthdays, anniversaries and major milestones. Some ($n = 3$) noted that they travelled to attend sport events. The most frequently reported travel activities were nature-based activities ($n = 25$) such as going to beaches or mountains, going to national parks, hiking, camping, finishing and skiing. When asked to talk about the impacts of family trips on the individuals and the family units, the overarching theme in participants' sentiments was hedonic well-being, with some indicating detachment and mastery, which are eudaimonic elements.

Hedonic well-being

Positive emotions and relationships were the most prevalent aspects of well-being in interviewees' narratives. Participants reported that they experienced a range of emotions, both positive and negative, on their family trips; however, the frequency and extensiveness (how many times something was mentioned and by how many participants) of positive emotion codes considerably outweighed the negative emotions; also, while interpersonal conflict and frustration were noted as stressors during family trips, interviewees' narratives showed that improved relationships with nuclear and extended family was a benefit of family trips. All in all, it was construed that, despite having ups and downs, family trips had made positive contributions to individual and collective family well-being.

Positive emotions

Joy, fun, excitement and relaxation were examples of positive emotions mentioned. For example, Participant 14 described his successful trips with his family as follows: '[I]t's something a little different, a little new, a little happiness, everybody gets along, assuming all the logistics are going well.' Participant 12 described her joyful experiences of the road trips with her children:

> I love road trips. I love driving, I love travelling. So do the older two. . . . So I would say that it used to be adventures and fun and I would have big maps and we would put stickers on the maps and then when we'd get to where the stickers were, the girls would get a book or something fun so they could read the map and have something to look forward to. . . . We just enjoy it.

Participant 18 described beach holidays in very similar terms: 'The beach is about relaxing. . . . This is my chance to get to be with [my husband and my son] and have no stress, well, less stress because I'm not trying to please anyone else. I think that's what vacation should be.'

As is apparent in participants' narratives, positive emotions were closely linked to connection and spending quality time with family, which leads us to another facet of well-being: relationships.

Relationships

Not surprisingly, positive relationships were highlighted numerous times as another hedonic well-being value accrued from family trips. Participants repeatedly stated that family trips provided them with opportunities to bond and spend quality time together, improve communication, strengthen family ties, create memories, repair damaged relationships, connect with extended family and family friends and develop patience and tolerance for others, all of which are indicators of what positive psychologists refer to as positive relationships, a pivotal aspect of well-being. For instance, Participant 17 shared that both she and her husband take their son on trips to reconnect with him and to do things he enjoys. She described those trips as 'very [son] focused, fun trips.' Participant 6 said, '[W]hen we are on a trip we have better communication. We have a book that's like 365 questions to ask your kids and they like it so much. . . . More time to communicate with your kids.' Similarly, Participant 11 explained that travelling offered her the opportunity to bond with her children. She stated:

> Oh, I love it cause that's the only time that I can actually get to spend time with them and get to see what kind of people my children actually are. Because daily interactions consist of just chores, getting to school on time, and things like that, and homework. But when you travel, you actually relax, and you get to hear about their hobbies. They start making comments about their friends; they start bickering with each other. You can see how stupid they can be. So, you get to know their personalities that I normally don't have access to because they express their personality with friends, not at home.

Participant 9 expressed the same sentiments about the need to remove oneself from everyday routine to be able to reconnect as a family: 'I think it gives us that quality time that we seek as a collective unit. . . . I think the whole family desires that.' Participant 1 used national holidays intentionally to see the extended family. She explained this is 'to spend quality time with our family again because we don't see them daily, weekly. At times I don't see my parents and my sister monthly so we take those opportunities.' Participant 9 had a similar opinion; she said:

> 'I think building relationships with people that we don't live close to. So, some of those cousins are [several states away] and I think it's important for the younger generations of our family to get to know each other so that when we're not around they still have each other.'

Participant 12 highlighted the importance of reconnecting with friends and maintaining those relationships: 'And then when we visit our friends, either they come up here or we go down there, it's just 'cause we are intentional about keeping that relationship going.'

In brief, families were 'able to focus on each other and spend time together and make memories' (Participant 15), while at the same time they learned 'to be patient and to have to be sitting in the car for nine or 10 hours. . . . It creates that overall bonding experience of dealing with each other in the car' (Participant 2). As is obvious in participants' statements, emotions and relationships as hedonic elements of well-being were mentioned in combination with learning and escape from routine, which presented eudaimonic well-being.

Eudaimonic well-being

Escape from routine and learning new things (i.e. detachment and mastery in well-being literature) showed up as eudaimonic aspects of well-being experienced in family trips, both of which were associated with positive emotions and relationships, as explained in the previous section.

Detachment

Having physical and mental detachment from work and the everyday demands of life has been recognised as one of the key benefits of travel in tourism scholarship. Specifically, in tourism and well-being studies, this is considered to be related to eudaimonia and is referred to as detachment. Participant 9 emphasised the need to escape from everyday routine to be able to reconnect as a family: 'Sometimes we need to just get away to make sure that [quality time with family] happens and get away from all the other stuff that's in our rigmarole.' Participant 6 brought up the importance of disconnecting from screens; she said, 'It's a way to encourage our kids to like being outside in nature because everything is so screen focused.' When asked to elaborate on that, she continued, '[I]t's the opportunity to get away from your day at home and you're able to connect without the added influences that you may not think about like dishes every day, laundry every day.' Similarly, Participant 17 explained that these trips are enjoyable because they bring a release from everyday responsibilities for the entire family:

> We don't have to worry about whether or not we're calling our parents enough or whether or not we're seeing certain friends enough or are we doing homework or our work. It's just an opportunity for us to spend some time just with each other and not have to focus on anything else. I think that's important.

Mastery

Most of the interviewees talked about family travel as an suitable context for learning. This involved learning interpersonal skills and learning about the outdoors, new destinations, other cultures, each other and the family's religion or traditions. Participant 8 travelled to new places to have diverse experiences: 'We pick somewhere different so we can have as many experiences as possible. They can experience new places, new cultures, new food, new activities. If they look back they will say we have all these memories.' Participant 5 briefly said, 'We like to take our kids to new places. We like to learn about other places, cultures.' Participant 13 explained that she likes taking her children to 'do some hiking and some camping, and go see beautiful places' because 'mostly with the kids we want them to have good experiences and see new places. . . . I want to have kids that enjoy the outdoors.' Similarly, Participant 14 wanted to expose his children to new parts of the country and 'to try to take the girls out west this summer, as a way to expose them to that part of the US'. Participant 15 took many trips with her family because she wanted her children to practice their foreign language skills and because it was important for her and her husband 'to have them see the world'. She explained:

> We all need to learn and see different things and learn a different rhythm of life. So, it's good to go anywhere where you're going to learn a different rhythm and be exposed to people who think differently than you do, and

how you do things. So that's kind of always part of our thing. We like to try new foods. We like to try activities. We like to see things.

Participant 4 wanted 'to have the kids experience good times, places, destinations'. Participant 17 also wanted to expose her son to 'different cultures, different people' and even believed that these experiences 'really helped make him a more outgoing person and someone who's more open to new experiences than he'd otherwise be'.

Passing on the family traditions to the children and introducing them to the places and experiences associated with the family's religion and/or culture by means of travel were also discussed by the participants. For example, Participant 14 mentioned that his wife and children went to Israel several times and that many of their travel experiences were related to their heritage. He described, 'I think an example would be going to [a destination five hours away], to celebrate a bar mitzvah . . . because the Jewish community in [our town] is zero, or maybe it's tiny.'

Similarly, Participant 16 shared the tradition of family reunions she was trying to pass on to her children:

We have a summer beach trip that we usually can make happen every summer. It's usually me, my mom, her siblings, their kids and grandkids, whoever's available. There's usually not like 20 of us, although we have gotten that big. . . . That's something that I even have been doing since I was a kid. My mom and her siblings would get together; we would all go to the beach or to an amusement park every summer. So, we're trying to keep that tradition going.

Discussion

The purpose of this study was to explore family travel and well-being from a positive psychology point of view. First, an extensive review of the literature on tourism and well-being was conducted to identify the eudaimonic and hedonic aspects of well-being in the extant tourism studies. Second, the identified elements were utilised as a guide to analyse a qualitative dataset aimed at assessing family travel and well-being. The analysis revealed two hedonic elements (positive emotions and relationships) and two eudaimonic elements (detachment and mastery). All four domains were found to be interconnected and interdependent in the participants' accounts.

Not surprisingly, hedonic elements of travel, such as relationships (bonding, connection, memory creation) and positive emotions (fun, enjoyment, happiness) appeared to be the strongest themes in this study. The main purpose of family trips for the majority was to strengthen relationships within the nuclear family unit or with the extended family. Most of the selected travel activities were aimed at improving relationships. Also, most of the positive emotions were accrued from spending time with the family without disruption of everyday life routines, and, as a result, enhanced relationships. This mirrored previous research that emphasised the importance of memory creation and bonding experiences with immediate and extended families (e.g. Shaw et al., 2008; Yun & Lehto, 2009). Previous research highlighted the importance of both everyday and unique shared experiences when it comes to family satisfaction (Zabriskie & McCormick, 2003), and this study supported the idea that travel may serve as one such unique family experience. Moreover, applying the hierarchical model of life satisfaction (Neal et al., 1999) to the findings of this study, we can suggest that family travel has the potential to enhance satisfaction with both family life and family leisure experiences (two domains of life) and, as a result, enhance overall life satisfaction.

The eudaimonic aspects were also related to relationships in that a great deal of the learning appeared to be about each other, which subsequently led to better relationships. Other areas of learning mentioned by the participants were different places, cultures, food, and people. The educational potential of family travels has previously been discussed in studies on travel benefits for families (e.g. Stone & Petrick, 2017; Wu et al., 2021). Also, detachment from routine during the trips was found to provide opportunities to 'focus on each other' and improve relationships.

The overlap in different domains of well-being supported previous work in the tourism and well-being space (e.g. Filo & Coghlan, 2016; Mirehie & Gibson, 2020a, 2020b). Indeed, Mirehie and Gibson (2020a) discussed that such overlap does not deny the fact that overall well-being consists of a range of eudaimonic or hedonic elements; rather, it may suggest that certain elements are more prevalent in specific contexts, noting the sensitivity of the well-being constructs to the context and the need for comparative studies to develop reliable measurement tools specific to tourism and well-being. Here, in the context of family travel, while four domains appeared in the participants' accounts, relationships appeared to be the focal point.

Although families still represent most leisure travellers within the US, travel experiences are still often an inaccessible luxury for many (Ambrosio, 2019). According to the US Family Travel Survey, the major limitations that stopped families from travelling were inability to afford family trips/holidays, too many other demands on the family budget and difficulty taking time off from work even in cases when paid leave days were available (Minnaert, 2021). Travel restrictions due to the COVID-19 pandemic exacerbated such long-standing concerns over the accessibility of travel for many families. Also, given the pandemic induced decline in mental health and well-being (Center for Disease Control and Prevention, 2021), we suggest that travel experiences can be important experiences for families. Thus, it is of great importance to ensure that all families can experience the benefits of such experiences.

To conclude, while assessing the downsides of family travel was outside of the scope of this chapter, it is important to recognise that, despite all of the benefits, family travel may bring stress, disruption of routines and conflict between family members (Kennedy-Eden & Gretzel, 2016; Rosenblatt & Russell, 1975). Thus, it is crucial to not create an unrealistically positive image of family travel experiences. We encourage the reader to avoid overestimating the potential benefits of family travel since multiple factors can affect such experiences (Rosenblatt & Russell, 1975) and many of those factors are outside of family control.

References

Ambrosio, R. (2019) Families treasure vacation time. Family Travel Association. See https://familytravel.org/2019-family-travel-survey-results-are-in. (Accessed December 7, 2021)

Andrews, F.M. and Withey, S.B. (1976) *Social Indicators of Well-Being: America's Perception of Life Quality*. New York: Plenum.

Braun, V. and Clarke, V. (2012) Thematic analysis. In H. Cooper, P.M. Camic, D.L. Long, A.T. Panter, D.E. Rindskopf and K.J. Sher (eds) *APA Handbook of Research Methods in Psychology, Vol. 2: Research Designs: Quantitative, Qualitative, Neuropsychological, and Biological*. Washington, DC: American Psychological Association.

Center for Disease Control and Prevention (2021) Coping with stress. See www.cdc.gov/coronavirus/2019-ncov/daily-life-coping/managing-stress-anxiety.html. (Accessed December 6 2021)

Charmaz, K. (2006) *Constructing Grounded Theory: A Practical Guide through Qualitative Analysis*. London: SAGE.

Chen, C.C., Huang, W.J. and Petrick, J.F. (2016) Holiday recovery experiences, tourism satisfaction and life satisfaction – Is there a relationship? *Tourism Management* 53, 140–147.

De Bloom, J., Geurts, S.A., Taris, T.W., Sonnentag, S., de Weerth, C. and Kompier, M.A. (2010) Effects of vacation from work on health and well-being: Lots of fun, quickly gone. *Work and Stress* 24 (2), 196–216.

Diener, E. (1984) Subjective well-being. *Psychological Bulletin* 95 (3), 542–575.

Diener, E. and Seligman, M.E. (2002) Very happy people. *Psychological Science* 13 (1), 81–84.

Diener, E.D., Emmons, R.A., Larsen, R.J. and Griffin, S. (1985) The satisfaction with life scale. *Journal of Personality Assessment* 49 (1), 71–75.

Dolnicar, S., Yanamandram, V. and Cliff, K. (2012) The contribution of vacations to quality of life. *Annals of Tourism Research* 39 (1), 59–83.

Durko, A.M. and Petrick, J.F. (2016) Travel as relationship therapy: Examining the effect of vacation satisfaction applied to the investment model. *Journal of Travel Research* 55 (7), 904–918.

Fennell, D.A. (2009) The nature of pleasure in pleasure travel. *Tourism Recreation Research* 34 (2), 123–134.

Fennell, D.A. (2018) On tourism, pleasure and the *summum bonum*. *Journal of Ecotourism* 17 (4), 383–400.

Filo, K. and Coghlan, A. (2016) Exploring the positive psychology domains of well-being activated through charity sport event experiences. *Event Management* 20 (2), 181–199.

Fredrickson, B.L. (2001) The role of positive emotions in positive psychology. The broaden-and-build theory of positive emotions. *American Psychologist* 56 (3), 218–226.

Fritz, C. and Sonnentag, S. (2006) Recovery, well-being, and performance-related outcomes: The role of workload and vacation experiences. *Journal of Applied Psychology* 91 (4), 936–945

Gao, M., Havitz, M.E. and Potwarka, L.R. (2020) Exploring the influence of family holiday travel on the subjective well-being of Chinese adolescents. *Journal of China Tourism Research* 16 (1), 45–61.

Gilbert, D. and Abdullah, J. (2004) Holidaytaking and the sense of well-being. *Annals of Tourism Research* 31 (1), 103–121.

Hill, B. (2000) Trends in family travel. Paper presented at the Fifth Outdoor Recreation and Tourism Trends Symposium, East Lansing, MI, September.

Institute for Social and Economic Research (2009) British household panel survey, wave 18. Colchester, Essex: The University of Essex.

Kammann, R. and Flett, R. (1983) Affectometer 2: A scale to measure current level of general happiness. *Australian Journal of Psychology* 35 (2), 259–265.

Kennedy-Eden, H. and Gretzel, U. (2016) Modern vacations – modern families: New meanings and structures of family vacations. *Annals of Leisure Research* 19 (4), 461–478.

Kluin, J.Y. and Lehto, X.Y. (2012) Measuring family reunion travel motivations. *Annals of Tourism Research* 39 (2), 820–841.

Kozak, M. (2010) Holiday taking decisions – The role of spouses. *Tourism Management* 31 (4), 489–494.

Kozak, M. and Duman, T. (2012) Family members and vacation satisfaction: Proposal of a conceptual framework. *International Journal of Tourism Research* 14 (2), 192–204.

Kozma, A. and Stones, M.J. (1980) The measurement of happiness: Development of the Memorial University of Newfoundland Scale of Happiness (MUNSH). *Journal of Gerontology* 35 (6), 906–912.

Kozma, A. and Stones, M.J. (1983) Predictors of happiness. *Journal of Gerontology* 38 (5), 626–628.

Lehto, X.Y., Choi, S., Lin, Y.-C. and MacDermid, S.M. (2009) Vacation and family functioning. *Annals of Tourism Research* 36 (3), 459–479.

Lehto, Y., Lin, Y.C., Chen, Y. and Choi, S. (2012) Family vacation activities and family cohesion. *Journal of Travel and Tourism Marketing* 29 (8), 835–850.

Lyubomirsky, S. and Lepper, H.S. (1999) A measure of subjective happiness: Preliminary reliability and construct validation. *Social Indicators Research* 46 (2), 137–155.

Madrigal, R. (2006) Measuring the multidimensional nature of sporting event performance consumption. *Journal of Leisure Research* 38(3), 267–292.

Minneart, l. (2021) *U S Family Travel Survey* New York Family Travel Association Mikke, M.V. and Stilling Blichfeldt, B. (2015) 'We have not seen the kids for hours': The case of family holidays and free-range children. *Annals of Leisure Research* 18 (2), 252–271.

Mirehie, M. and Gibson, H.J. (2020a) The relationship between female snow-sport tourists' travel behaviors and well-being. *Tourism Management Perspectives* 33, 100613.

Mirehie, M. and Gibson, H.J. (2020b) Women's participation in snow-sports and sense of well-being: A positive psychology approach. *Journal of Leisure Research* 51 (4), 397–415.

Moal-Ulvoas, G. (2017) Positive emotions and spirituality in older travelers. *Annals of Tourism Research* 66, 151–158.

Neal, J.D., Sirgy, M.J. and Uysal, M. (1999) The role of satisfaction with leisure travel/tourism services and experience in satisfaction with leisure life and overall life. *Journal of Business Research* 44 (3), 153–163.

Neal, J.D., Uysal, M. and Sirgy, M.J. (2007) The effect of tourism services on travelers' quality of life. *Journal of Travel Research* 46 (2), 154–163.

New Economics Foundation (NEF) (2009) National accounts of well-being: Bringing real wealth onto the balance sheet. London: NEF. https://neweconomics.org/2009/01/national-accounts-wellbeing

Newman, D.B., Tay, L. and Diener, E. (2014) Leisure and subjective well-being: A model of psychological mechanisms as mediating factors. *Journal of Happiness Studies* 15 (3), 555–578.

Oliver, R.L. (2010) *Satisfaction: A Behavioral Perspective on the Consumer* (2nd edn). Armonk: M.E. Sharpe.
Peterson, C., Park, N. and Seligman, M.E. (2005) Orientations to happiness and life satisfaction: The full life versus the empty life. *Journal of Happiness Studies* 6 (1), 25–41.
Presser, H. (2000) Nonstandard work schedules and marital instability. *Journal of Marriage and the Family* 62, 93–110.
Rosenblatt, P.C. and Russell, M.G. (1975) The social psychology of potential problems in family vacation travel. *Family Coordinator*, 24 (2) 209–215.
Ryff, C.D. and Keyes, C.L.M. (1995) The structure of psychological well-being revisited. *Journal of Personality and Social Psychology* 69 (4), 719.
Shaw, S.M. and Dawson, D. (2001) Purposive leisure: Examining parental discourses on family activities. *Leisure Sciences* 23 (4), 217–231.
Shaw, S.M., Havitz, M.E. and Delemere, F.M. (2008) 'I decided to invest in my kids' memories': Family vacations, memories, and the social construction of the family. *Tourism Culture & Communication* 8 (1), 13–26.
Seligman, M.E. (2002) *Authentic Happiness: Using the New Positive Psychology to Realize Your Potential for Lasting Fulfillment*. New York: Free Press.
Seligman, M.E. (2004) Can happiness be taught? *Daedalus* 133 (2), 80–87. DOI: 10.1162/001152604323049424.
Seligman, M.E. (2011) *Flourish: A Visionary New Understanding of Happiness and Well-Being*. New York: Free Press.
Seligman, M.E.P. and Csikszentmihalyi, M. (2000) Positive psychology: An introduction. *American Psychologist* 55 (1), 5–14. https://doi.org/10.1037/0003-066X.55.1.5.
Smith, W.W., Pitts, R.E., Litvin, S.W. and Agrawal, D. (2017) Exploring the length and complexity of couples travel decision making. *Cornell Hospitality Quarterly* 58 (4), 387–392.
Sonnentag, S. and Fritz, C. (2007) The Recovery Experience Questionnaire: Development and validation of a measure for assessing recuperation and unwinding from work. *Journal of Occupational Health Psychology* 12 (3), 204.
Stone, M.J. and Petrick, J.F. (2017) Exploring learning outcomes of domestic travel experiences through mothers' voices. *Tourism Review International* 21 (1), 17–30.
Strauss-Blasche, G., Ekmekcioglu, C. and Marktl, W. (2000) Does vacation enable recuperation? Changes in well-being associated with time away from work. *Occupational Medicine* 50 (3), 167–172.
Watson, D., Clark, L.A. and Tellegen, A. (1988) Development and validation of brief measures of positive and negative affect: The PANAS scales. *Journal of Personality and Social Psychology* 54 (6), 1063–1070.
West, P.C. and Merriam Jr, L.C. (2009) Outdoor recreation and family cohesiveness: A research approach. *Journal of Leisure Research* 41 (3), 351–359.
WHOQOL Group (1995) The World Health Organization quality of life assessment (WHOQOL): Position paper from the World Health Organization. *Social Science & Medicine* 41 (10), 1403–1409.
Wu, W., Kirillova, K. and Lehto, X. (2021) Learning in family travel: What, how, and from whom? *Journal of Travel & Tourism Marketing* 38 (1), 44–57.
Yun, J. and Lehto, X.Y. (2009) Motives and patterns of family reunion travel. *Journal of Quality Assurance in Hospitality and Tourism* 10 (4), 279–300.
Zabriskie, R.B. and McCormick, B.P. (2003) Parent and child perspectives of family leisure involvement and satisfaction with family life. *Journal of Leisure Research* 35 (2), 163–189.

8 The Trinidad Carnival and the Promotion of *Joie de Vivre*

Johnny Coomansingh

Introduction

Notwithstanding the fact that there are carnival celebrations all over the world, the Trinidad Carnival is peddled as the one that is the most imitated, the most copied, and the most photographed (Mason, 1998). 'Many countries have a carnival, but carnival is Trinidad – and Trinidad is carnival. . . . Brazil may boast larger attendances but Trinidad is its true spiritual home. It has hosted the carnival for longer, and in a far more expressive form' (Mason, 1998: 7). The Trinidad Carnival has been described in these words:

> Carnival occupies Trinidad's attention as nothing else, if you are not taking part you are planning to take part, and if you are not talking about taking part then you are talking about what happened last year. . . . In some countries carnival is a diversion from the troubles of life; in Trinidad it sometimes seems as if life is a diversion from carnival. (Mason, 1998: 16)

From an existential viewpoint, the pre-Lenten Trinidad Carnival is a time outside of time with portrayals of gustatory excesses combined with raucous and ribald revelry. The celebration serves as an annual attraction for tourists from all over the world. Most of the tourists come from Europe, North America, the Caribbean region and Japan. Why do tourists make the annual journey to Trinidad for its carnival? It is possible that there is an innate desire to be happy, to be joyful, and to express oneself, and perhaps carnival is the avenue that provides them with such opportunities. Table 8.1 gives an indication of the number of tourists that visited Trinidad and Tobago for the carnival season 2016–2020.

There are many calypsonians in Trinidad and Tobago who are famous for their calypsos about female tourists who come to Trinidad for the carnival season. Dr Francisco Slinger (sobriquet: The Mighty Sparrow), the Calypso King of the World (Kwamdela, 2006), has sung several of these songs. Not only did Sparrow sing about 'Mrs White', the tourist who

Table 8.1 Visitors to Trinidad and Tobago and estimated expenditure for the carnival season 2016–2020

Year	2016	2017	2018	2019	2020
Total visitors	35,483	37,448	33,873	35,560	37,861
Estimated expenditure (TTD)	340,530,351	334,897,464	317,796,486	388,493,000	458,155,961

Source: Central Statistical Office (n.d.).

came to meet him one carnival in Port of Spain, but Aldwin Roberts (sobriquet: Lord Kitchener) also made his mark with the calypso 'Miss Tourist' (Grant, 1993). Here is an excerpt from the calypso:

> Ah Tourist dame, I met her the night she came
> Well she curiously asking about my country
> She said I heard about bacchanal and Trinidad Carnival
> So I come to jump in the fun
> [for rest of lyrics see https://genius.com/Eddy-grant-miss-tourist-lyrics]

Of course, this is what the tourists come here for; this is the epitome of *joie de vivre*, this exuberant enjoyment of life they desire to experience in the celebration of the Trinidad Carnival, to fete and party with calypso and steelband.

Although the people of Trinidad are well known to possess the carnival mentality – a live-for-the-moment attitude (Green, 1998) – in terms of power of place (Berdoulay, 1989; Robinson, 1989), the annual carnival ritual reveals the true sinew of a people to express their freedom, to fete, party, and to enjoy themselves as though there is no tomorrow. In essence, 'the Trinidad Carnival is in fact a kind of superorganic entity (Zelinsky, 1992; Mitchell, 2000), the basic value system, the glue that directly or indirectly holds the society together' (Coomansingh, 2011: 120). Trinidad's identity and sense of place is tied firmly to the carnival. As a corollary to this carnival mentality, Trinidad has also been labelled with the 'any time is Trinidad time' lifestyle (Birth, 1999). In their daily rounds, many citizens do not recognise time management as a matter for serious concern. Calypsonian Bunji Garlin referred to the Trinidad Carnival as a deep sea (Christopher, 2021). In other words, no one can really fathom or comprehend the depth of what Trinidad Carnival truly is. Nonetheless, the pre-Lenten Trinidad Carnival did not just happen overnight. The annual event came from somewhere with all its rigour and revelry.

Carnival Origins

Though it is only speculation, an explanation is given here about the origins of the pre-Lenten carnival as practiced in Trinidad. History suggests that in ancient times a type of rebellious was exhibited in public spaces during carnival celebrations. Today, on the public streets of Trinidad, the display of certain lewd, sensuous and vile behaviours during the festivities implies that such behaviours were copied and developed to suit the participants. Over time, behaviours, attitudes and costumery metamorphosed into a fusion of the sacred and the profane. Some of the more resilient icons have been retained while others have been sanitised or altogether lost.

Celebrated as a period of gustatory excess and almost uncontrolled behaviour, the term *carnival* from the Latin *carne levare* or *carnelavarium* means to take away fat or meat (Gilmore, 1988). Noisy merrymaking, riotous revelry and raucous and ribald behaviours during street processions precede Ash Wednesday. According to Gill (1997), carnival festivities historically signified a period of feasting and revelry that commenced on Twelfth Night or Epiphany and ended on Shrove Tuesday, or Mardi Gras. Mardi Gras, which means Fat Tuesday, was supposed to leave the faithful of the church in good shape to face the self-denying ordinances of Lent.

Carnival of the Egyptians, Greeks and Romans

It is also probable that the Egyptians during the 12th dynasty were the first recorded culture that celebrated carnival. Carnival was described by Smart and Nehusi (2000) as the ultimate pan-African festival, a type of pre-Lenten celebration that had its genesis in the Nile Valley in Kemet (ancient Egypt). This carnival was celebrated near to the very dawn of human history. The only humans that were in existence at that time were the 'Afrikans' and they were the ones who annually staged the Worsirian Festival or mystery play, the mother of all festivals (Smart & Nehusi, 2000).

During that era in Egypt, five of the 365 days of the year were set aside to recreate or restore harmony to their relationship with the deities of the cosmos. This period, in which the Egyptians would chant ribald songs, drink brew, and carouse, was known as a time outside of time. Along with carousing, the chanting of ribald songs, and the drinking of fermented brew, there would be torch parades with women revellers holding aloft giant erect phalluses. This theatrical spectacle was actually a re-enactment of the passion that existed between Isis and her husband/brother Osiris, who incidentally was the Egyptians' god of rebirth (Coomansingh, 2019). In other words, the peoples of Europe were celebrating the carnival that the Egyptians transmitted to them centuries ago. De Blij and Muller (2002) also allude to the idea that Africa was the nursery of civilisation and that all races emerged from Africa.

Both the Greeks and Romans celebrated a type of carnival, but the origins of such a carnival are probably lost in the mists of time (Gilmore, 1998); nevertheless, it is apparent that the Greeks had a concept of carnival in Europe by 1100 BC. As theorised by Sautman (1982), carnival was declared to be perhaps the most ancient of all Western pagan festivals observed today. Marking the onset of the Christian fast, or Lent, is the medieval European carnival. This carnival was probably related to Roman holidays and/or festivals (Gilmore, 1998). Supposedly connected to carnival events were the gods of classical antiquity, Bacchanalia, Lupercalia and Saturnalia.

Bacchus and Dionysius, the gods of wine and debauchery, were the gods to which the Greeks and Romans respectively paid homage. Over time, Lupercalia came to be associated with the festivals, and this was evidenced in the portrayals of acts of licentiousness, drunkenness, and debauchery. Around the middle of February, the god Pan, or the faun, was feted in the festivals of Lupercalia. The Roman Saturnalia observed the god Saturn in mid-December. A satyr-like figure became the king for the day in the festival celebrating the King of Saturnalia.

In the French village of Cournonterral during the 14th century there was an aggressive flaunting of putrefaction during its carnival. This 'celebration' signified the bridging of the gap between life and death. For the occasion, Cournonterral has a song: 'We are of blood and wine, the more it rains, the thicker the mud – we are happy in our filth' (Sautman, 1982: 67). Besides the wine sediment normally used for the occasion, manure, compost heaps, blood, tripe, decomposed animal corpses, and the contents of stables and lavatories were at one time employed in the celebration. During this period, certain aspects of carnival as practiced in parts of Europe involved groups of men in southern France smearing their entire bodies with dark substances. Parading sometimes entirely naked, they would chase women with a long stick held like a phallus while singing, 'We want to fornicate' (Sautman, 1982: 22). Nevertheless, regardless of its origins, the carnival with all its flair, grandeur, and decadence arrived in Trinidad with the French.

The French plantocracy came with permission from the Spanish crown. In 1783, the king of Spain published a *cédula de población* (cedula of population) that attracted French Roman Catholics from other Caribbean islands to settle on lands in Trinidad. As is normal, people come with their cultural baggage (Zelinsky, 1992). Without any doubt, therefore, it was the French with their *fêtes champêtres* who brought the carnival to Trinidad (Gill, 1994). Before the arrival of the French and their enslaved Africans, no form of any carnival was practiced on the island (Cowley, 1996; Mason, 1998). Carnival in Trinidad, after the French arrival in 1783, was an affair associated with elaborate balls, house-to-house visitation, street promenading in carriages, and masking and costuming (Hill, 1972). Included also in the carnival were playwrights and gentlemen actors.

During the two days of carnival in Trinidad, Carnival Monday (Lundi Gras) and Carnival Tuesday (Mardi Gras), carnival celebrants advance opportunities for fresh forms of confrontation and exchange (Koningsbruggen, 1997), which, in one way or another, resonate with the struggles and battles between slave owners and enslaved Africans (Hill, 1972). Since time immemorial, regardless of carnival rules and regulations sanctioned by government authorities, confrontation is still extant on the landscape at every carnival. Although Trinidad is considered to be one of the wealthier states in the Caribbean region, 'the tiger in the sea of pussycats' (Rohter, 1998), there are shades of poverty that convey the idea that there is serious social differentiation, much of which is noticeable on some of the landscapes adjacent to the cities. A great deal of the rebellion and resistance to law and order is transmitted via bacchanalian behaviours in the masquerade bands.

Carnival as an 'Inversion of Normal Social Order'

According to Bakhtin's sociological view, carnival is an activity that does not entertain or seem to be cognizant of spatial boundaries because, during any carnival, people live only according to its 'laws', the laws of carnivalesque freedom (Bakhtin, 1968). For all intents and purposes, carnival is hedonistic – the pure pursuit of individual pleasure. Nevertheless, Mitchell (2000) considers carnival as the dethronement of the sacred, a movement of resistance, an act of transgression, an inversion of normal social order, a license in ritual, and a means of releasing steam. 'Carnivals, fairs and everyday life are a powerful set of tools for subordinated culture that constantly undermine the presumptions of elite culture. The inversion of symbolic domains of "high" and "low", for instance, pokes fun at the establishment and irritates the agents of culture' (Mitchell, 2000: 161).

Carnival is viewed by Sampath (1997) as a sensory relief for those who are most blinded by a spectrum of supposedly civilised white colonialism and its respectability. Yet it is at the same time the heart and soul of true cultural freedom and a reaffirmation and celebration of positive identity for those who are most affected by colonialism (Coomansingh, 2002). Carnival beckons revellers to bacchanalian gatherings; it is a literal exercise in sensuality and harmony but yet there is bedlam, clamouring vendors selling baubles, leering faces, frightening masks, dancing, shouting and whistling, all engulfed in aggressive, pulsating rhythms. Carnival has also been depicted as a licentious mass festival involving rituals of conflict and rebellion; episodes of ridicule, threat and rivalry; temporary suspension of the rules of social order; and an aggressive display of unpredictable behaviour (Linger, 1992).

As mentioned before, pre-Lenten carnivals seem to be a composite of the sacred and the profane. In Trinidad, there is definitely some form of cultural syncretism between the carnival and the Roman Catholic Church. It is a fact that an intimate relationship exists between the church and the carnival (Green, 1998). In 1995, although the Pentecostal Church was furious over the use of the term 'Hallelujah' as the name of a carnival band, Father Clyde Harvey of the Roman Catholic Church defended Peter Minshall, the band leader. Instead of rebuking him, Harvey praised Minshall for his efforts to please God. It is well known that the Pentecostal Church in Trinidad reveres the term 'hallelujah' and called on the government to prevent the use of the term. The Roman Catholic Church has supported and defended the Trinidad Carnival throughout its history because the celebration is basically a Roman Catholic tradition (Coomansingh, 2002).

Calypso and Steelband

During the carnival season, the dialect of carnival, influenced by the calypsos playing on the airwaves, takes over the airwaves; double entendre and sexual innuendo pervade the atmosphere (Dewitt, 1993) immediately after Christmas Day. For example, words such as *jam* and *wine* take on radically different meanings. *Jam* does not have anything to do with a sweet fruity paste, and *wine* has nothing to do with an alcoholic beverage (Green, 1998). Every year there are new calypsos. No old calypso is rehashed for use during any carnival. Many of the calypsos arrive on the carnival landscape with the aim of persuading people to do things and to become active – to excite themselves and participate in the music.

Many calypsonians create an atmosphere where celebrants get into an uncontrollable ecstatic frenzy, generating acts of 'wining' (a type of sensual dance). Most women who participate in the Trinidad Carnival provoke the sexual appetites of the opposite sex through their wining exhibitions (Koningsbruggen, 1997). Sometimes expressing oneself during the carnival becomes an extreme act. One female celebrant, a schoolteacher by profession, having been hypnotised by the words of the calypsonian Austin Lyons (sobriquet: Blue Boy) with his 'Get Something and Wave' calypso, pulled off her underwear and waved them in the air while wining simultaneously, much to the amusement of some and the astonishment of others (Coomansingh, 2002). This is what the Trinidad Carnival evokes. People lose themselves and forget their surroundings. It is a moment in time when the celebrant alone has the power to make his or her life wonderful, regardless of who is looking.

Recently, another calypsonian, Hollis Mapp (sobriquet: Mr Killa), with his 2019 rendition 'Run Wid It', actually influenced people to crazily grab things and run with them. This form of activity is also a factor in the excitement and exuberance of the carnival experience. It's on record that the activity generated by this song prompted mixed views. According to Kong Soo and Dixon (2019), 'On hearing Hollis "Mr Killa" Mapp's song *Run Wid It*, patrons have been picking up tents, coolers, chairs, barricades, furniture, corn soup, other people's property, and even women without their permission.'

The music of the calypso is typically interpreted by the steelpan (Coomansingh, 2005). No other medium seems to be satisfactory for the presentation of calypso music on carnival day, especially when it comes to the Road March. The Road March is a calypso that is played more than any other calypso during Lundi Gras and Mardi Gras. Many calypsonians vie for this title every year. The greatest Road March King of all time is Aldwin Roberts (sobriquet: Lord Kitchener), otherwise known as the Grandmaster of Calypso. In one of his calypsos, he sang, 'The road make to walk on carnival day. Ah doh want to talk but ah have to say, any steelband man only venture to break this band, is a long funeral to the general hospital.' His lyrics go to show that on carnival day nobody should interfere with anybody, because the road make to walk on carnival day. This phrase conveys 'I claim this space', and, as Linda McCartha Monica Sandy-Lewis (sobriquet: Calypso Rose) said, 'Don't touch me!' Without the renditions of calypsos, it would be a lacklustre carnival, and without the sound of the steelpan orchestras, it would be a very dead carnival.

The creation of the steelpan in Trinidad must be understood in terms of the island's 'history of colonialism and decolonization, class hegemony and resistance, and ethnic diversity and creolization' (Stuempfle, 1995: 11). The steelpan was used by the subordinated culture to actually 'undermine the presumptions of elite culture' (Mitchell, 2000: 161). Complementing the tamboo bamboo were the dances: Limbo, Bongo and an African twist on the French Belle Aire dance (Blake, 1995; Cowley, 1996; Koningsbruggen, 1997; Mason, 1998; Stuempfle, 1995). This simple group of makeshift metal objects (instruments) and the bottle and spoon combination laid the foundation for the creation of steelpan as it is known today (Blake, 1995). Although the class of people involved in the creation of the steelpan were to some extent, ragged, rough, rebellious, raucous, and ribald, they continued with their revelry and resistance to satisfy the needs of the Trinidad Carnival.

This element of Trinidad Carnival has gradually given recognition to the lower classes or what is referred to as the 'jamet' (from the term *diametre* or *diamet* – a woman who lives below the level of respectability) class of people. Initially the steelpan was associated with street gangs and the backyard life of the city of Port of Spain (Coomansingh, 2002). To be a pan man, one who plays the steelpan, lives for it, and inhabits or loiters in the panyard, was anathema prior to the 1950s (Koningsbruggen, 1997). A pan man had very little respect from the populace.

Not only were pan men sidelined and regarded as not fit for existence in civil society, their followers from the shanties of Port of Spain were also relegated to the same class. As historical records show, between 1860 and 1880, carnival in the capital city, Port of Spain, and the second largest city, San Fernando, became a type of war with village rival groups of 'badjohns' (violent male revellers) and jamets waging war with each other and also the police (Cowley, 1996; Koningsbruggen, 1997). Sticks, stones, and bottles were some of the weapons used during the 1860s in the rivalry episodes on carnival day. The once 'peaceful' carnival was now entirely overtaken by the jamets (Koningsbruggen, 1997).

The violent behaviours exhibited by the badjohns and jamets are but the leftovers of European-type carnivals. Violence is another common characteristic of pre-Lenten carnivals. Carnival in the town of Penia, Italy, for example, is a revival of the pre–World War I Masquerade of the Plough and Masquerade of the Wedding Party (Poppi, 1991). The 'Ugly Masks' (*mescres a burt*) and 'Handsome Masks' (*mescres a bel*) formed the bulk of the pageant, which depicted opposing forms of behaviour. A greater degree of ritual license surrounded the Ugly Masks, and anything 'ugly' could be expected from a group of Ugly Masks. Under the cover of the masks, violent eruptions could ensue, especially when parties from different villages collided. Encounters of this type were sometimes fatal (Poppi, 1991). During this time of year the 'evil' of society would emerge (Smart & Nehusi, 2000). But, despite the hovering violence, the carnival must be staged, and in Trinidad people are bound to play mas.

The Masquerade (Mas) and the Trinidad Carnival

In personal communication with Kenrick 'Watchie' Watch, a traditional mas designer, mas creator, and band leader, he mentioned that there are two kinds of masqueraders: Traditional Mas and Conventional Mas. Watchie spoke about traditional mas characters as carrying with them some kind of historical import or baggage from the carnivals of yesteryear. One thing that must be noted is that, historically, almost all masqueraders were fully clad. Note also that the carnival has been overtaken by women. Thousands of bikini-clad women now take to the streets clad in skimpy bikinis with just a few sequins and feathers covering the most private parts of their anatomy. At one time, traditional mas characters such as Dame Lorraine, Clowns, Pierrot Grenade, Bat Mas, Devils, Dragons, Gorillas, Jab-Jab, Warrahoons, Moko Jumbie, Apache (Red Indian), Gatka, Midnight Robbers, Negue Jardin, Baby Doll, Bookman, Burrokeet, War Mas, Sailor Mas (Coomansingh, 2019), and others were noticed to be fading on the carnival landscape.

From personal observation, there was one particular character, a man nicknamed 'Piss', who played Lady Mas. Dressed as a lady in *her* finest lace, Piss danced like a mesmerised Sufi in the middle of Cunapo as if there were no tomorrow. He found a way to express himself in place for everyone to see and admire his happiness, freedom, and joy. Today, there is an upsurge in the presentation of traditional mas, and Watchie insists that the eastern town of Sangre Grande is the home of traditional mas. He said that he has played Mud Mas or Dutty Mas for as long as he could remember. 'There is a feeling that comes over me on Carnival Sunday *Dimanche Gras* that is almost uncontrollable. I don't know what it is. I can't explain it. All I know is that I have to go down to Cunapo (Central Business District/Sangre Grande) early on Carnival Monday morning to get involved in the Jouvay,' he said.

It would seem that Jouvay, especially the Mud Mas, a carryover from ancient European carnivals, has become an almost essential part of the Trinidad Carnival. Both citizens and tourists participate in this 'ritual'. After a bath of watery mud from head to toe under the cover of darkness, lewd and sensual acts commence. The Mud Mas is a type of baptism or burial from which revellers emerge later to blossom into conventional mas or 'pretty mas' the following day. Most people, inebriated to the max, are tired and weary by this point, but this does not stop them. People participating in Jouvay party on the streets from 2am until about 11am. They keep joyfully chipping to the music from the big trucks equipped with music and speaker boxes that blast calypso and soca music, vibrating everything around them. After the revellers can take it no more, they go home to clean up and rest. Nevertheless, they will soon be back on the streets to participate in what Watchie described as 'street theatre', the exhibitions of traditional mas and conventional mas. Conventional mas normally parades on Carnival Tuesday with an array of sequins, feathers and women clad in skimpy bikinis.

And in amongst all this chaos, bedlam, confusion, noise and commotion, something is imperceptibly happening to participants: a creeping sensation of the need to wine. The wining commences at Jouvay as if to say, 'This is my day. I waited all year to play mihself, and ah doh care. I want to free up. I want to express myself. I need to feel wonderful about myself'. In this activity, after being pent up all year, a rebellious type of happiness takes over the celebrant. The Trinidad Carnival does this to people in every village and every town. Churchgoer or no churchgoer, a powerful gyration of the hips ensues and the entire body goes into a sensual and sexual dance, swivelling from side to side and even motioning at intervals in backward and forward forays termed as juking (Yawching, 2002). Koningsbruggen (1997) mentions that the custodians of decent morality constantly criticise the activity. The act of wining is summed up as:

> the rotating of the hips and moving them up and down in an exaggerated manner which is sexually suggestive, and leaning further and further forward or backwards while bending at the knees. . . . [M]en and women do this dance in pairs or in a row, which may number eight or ten dancers. (Koningsbruggen 1997: 82)

But isn't wining a part of the carnival? In Trinidad there is a statement, 'Yuh cyar play mas an' 'fraid powder' (Mason, 1998: 90). In other words, you can't be in something and not be a part of it. To stop the wining would be an injustice to the many revellers who want to 'free up' themselves for freeing up's sake; many of them – especially women, according to Koningsbruggen (1997) – wine not because they want sex but as a means through which they can express their sexuality.

The whole pattern of obscenity has been a part of the Trinidad Carnival for several generations. In personal communication with Nigel Mohammed, he described the wining dance as similar to a mating dance, because the stage is set with all the necessary elements converging: the people (men and women), the place, the beverages and other inebriating substances, and of course the propensity to wine. He said that the carnival was a forum for people to come together to enjoy and express their desires. There are so many scenes to any landscape that a person can find through the beholding eyes (Meinig, 1979). As Wilkes (2002) affirmed, 'And when they start to fling the thing. And Rocky hold them from behind. Who could resist temptation. Just roll back and wine and wine?' It's amazing – more like astonishing – to see prissy bank workers, teachers, attorneys, and hoity-toity people of certain professions and walks of life freeing up and wining on carnival day and then, after the carnival, going to church to receive ashes on Ash Wednesday, to be blessed by a priest and crawl back into their various enclaves as if they exhibited nothing lewd and sensual. But who cares if such people feel wonderful and excited about their activities and accomplishments on carnival day? That is the Trinidad Carnival.

Conclusion

For over 200 years, a pre-Lenten carnival with all its chaos, confusion, rebellion, rancour, disorganised organisation, and rituals has found its place on this outcropping of rock in the Caribbean Sea, Trinidad. By accident or design, nature bequeathed this island with the honour of having a multi-racial and multi-religious population who, under slavery and indentureship, forged the Trinidad Carnival bit by bit and piece by piece. The Trinidad Carnival is now a celebration where in every nook and cranny, every town and village, participants including tourists can express their freedom, joy and happiness. The celebration becomes a moving, gyrating landscape with citizens and thousands of tourists who come to learn how and when to wine, how to perform onstage in a masquerade, and how to play the steelpan. Every year the tourists come for a new experience that they can take back to their homelands.

The Trinidad Carnival was formed through the synergistic nature of the people and the place. Robinson (1989: 157) explained, 'Place is synergistic; it is created and it creates; that place is constructed, destructed, and transformed by individuals, and/or higher level corporate groups within special cultural contexts'. With every year that passes something new is added to the carnival without erasing the old, which speaks to the morphology of landscape (Sauer, 1925). In this land of sun, sea and rum (Potter & Lloyd-Evans, 1997), carnival encompasses the emotions, the aspirations, the rebellion, the resistance, and the violence embedded in the people; the latter aspect, regretfully, has become a more sophisticated threat to human life by 'the people' of the carnival. Berdoulay (1989: 125) posits, 'The idea of place implies a meaningful portion of geographical space. Place involves meaning for the people who build it, live in it, visit it, or study it . . . in the sense that there is a special, emotional link between people and place'.

With their polyrhythmic and rebellious nature, these people resisted their oppressors in the satirical content of their calypsos, but they did not stop there. Although disobedient to their colonisers, they eventually created the steelpan instrument, the only symphonic percussion instrument created in the 21st century. The argument here is to illustrate that people, despite their struggles and hopelessness, have something inside them that they wish to release, to express, to share, and to commemorate. Nevertheless, apart from the violence that occurs in certain instances on carnival days, there are centripetal forces in the Trinidad Carnival that operate to categorise the masses. Some large bands tend to accommodate people of lighter skin colour while rejecting those whose skin contain more melanin. It would seem here that the upper echelons of society, the remnants of the French creoles, still want to remain aloof. This does not bode well for the Trinidad Carnival.

But what really is carnival? Is carnival just a moment in time where celebrants turn out in their droves to go against the rules of law and order? Is it a time when people decide to misbehave in their towns and villages? Is it an occasion on which to become sensuous and sexually liberated? Is it a moment to celebrate what the status quo describes as indecency? Is it a chance to express hidden desires and tastes, to give release to pent up feelings harboured all year? Could carnival with all its rebellion be described as a latent landscape in the mind of celebrants? Does the building of a metaphysical landscape spring into *joie de vivre* – the exuberant enjoyment of life – to become a literal, physical landscape in motion when the actual moment of celebration arrives? I am of the view that carnival is just that. Even from Ash Wednesday, people begin to plan how they will dress, how they will look, and how they will behave, how drunk they will get, how much they will wine, and on whom they will wine. Carnival occupies the mind of Trinidadians as nothing else. From the early age of 11, I learned that happiness is a matter for the individual. I understood that happiness is a clear conscience and the joy of just being alive. You and you alone have the power to make your life wonderful and nothing in the world does more to promote this happiness, joy and freedom than the Trinidad Carnival.

References

Bakhtin, M. (1968) *Rabelais and His World*. Cambridge, MA: MIT Press.
Berdoulay, V. (1989) Place, meaning and discourse in French language geography. In J.J. Agnew and J.S. Duncan (eds) *The Power of Place* (pp. 124–139). Boston: Unwin Hyman.
Birth, K. (1999) *Anytime is Trinidad Time*. Gainesville: University Press of Florida.
Blake, F. (1995) The Trinidad and Tobago steel pan: History and evolution. See www.seetobago.com/trinidad/pan/ref/tamboo2.htm (accessed May 2002).

Central Statistical Office (n.d.) Tourism statistics. See https://cso.gov.tt/subjects/travel-and-tourism/tourism-statistics (accessed May 2002).

Christopher, P. (2021) Carnival's data gap. *The Trinidad Guardian*. See www.guardian.co.tt (accessed May 2021).

Coomansingh, J. (2002) Trinidad Carnival as an amalgam of borrowed cultural elements. MA thesis (unpublished), Kansas State University.

Coomansingh, J. (2005) The commodification and distribution of the steelpan as a conflicted tourism resource. PhD dissertation (unpublished), Kansas State University.

Coomansingh, J. (2011) Social sustainability of tourism in a culture of sensuality, sexual freedom and violence: Trinidad and Tobago. In J. Carlsen and R. Butler (eds) *Island Tourism: Towards a Sustainable Perspective* (pp. 118–128). Wallingford: CABI.

Coomansingh, J. (2019) *An Understanding of the Trinidad Carnival: A Melange of Borrowed Cultural Elements*. Denver: Outskirts Press.

Cowley, J. (1996) *Carnival, Canboulay and Calypso*. Cambridge: Cambridge University Press.

De Blij, H.J. and Muller, P. (2002) *Geography, Realms, Regions, and Concepts*. Chichester: John Wiley and Sons.

Dewitt, D. (1993) *Callaloo, Calypso and Carnival: The Cuisine of Trinidad and Tobago*. Toronto: Crossing Press

Gill, J. (1997) The rise and fall of the French carnival. In *Lords of Misrule: Mardi Gras and the Politics of Race in New Orleans* (pp. 26–58). Jackson, MS: University Press of Mississippi.

Gill, M.L. (1994) Presence, identity and meaning in the Trinidad Carnival: An ethnography of schooling and festival. Doctoral thesis (unpublished), University of Wisconsin-Madison.

Gilmore, D.D. (1998) *Carnival and Culture: Sex, Symbol and Status in Spain*. New Haven Ct.: Harvard University Press.

Grant, E. (1993) Miss Tourist, *Soca Baptism*. See https://genius.com/Eddy-grant-miss-tourist-lyrics (accessed June 2022).

Green, G.L. (1998) Carnival and the politics of national identity in Trinidad and Tobago. PhD dissertation, New School for Social Research.

Hill, E. (1972) *The Trinidad Carnival: Mandate for a National Theater*. Austin: University of Texas Press.

Kong Soo, C. and Dixon, B. (2019) Mixed views over Mr Killa's song: Hit makes people take up anything and 'run with it'. *Trinidad Guardian*, February 16. See www.guardian.co.tt/news/mixed-views-over-mr-killas-song-6.2.782049.71f697cdee (accessed July 2021).

Koningsbruggen, P. (1997) *Trinidad Carnival: A Quest for Identity*. London: Macmillan.

Kwamdela, O. (2006) *Mighty Sparrow, Calypso King of the World*. Brooklyn: Kibo Books.

Linger, D.T. (1992) *Dangerous Encounters: Meanings of Violence in a Brazilian City*. Stanford: Stanford University Press.

Mason, P. (1998) *Bacchanal: The Carnival Culture of Trinidad*. Nottingham: Russell Press.

Meinig, D.W. (1979) The beholding eye. In D.W. Meinig (ed.) *The Interpretation of Ordinary Landscapes* (pp. 1–9) New York: Oxford University Press.

Mitchell, D. (2000) The dialectics of spectacle. In *Cultural Geography: A Critical Introduction* (pp. 129–147). Chichester: John Wiley and Sons.

Poppi, C. (1991) Building difference: The political economy of tradition in the Ladin Carnival of Val di Fassa. In J. Boissevain (ed.) *Revitalizing European Rituals* (pp. 112–155). London: Routledge.

Potter, R.B. and Lloyd-Evans, S. (1997) Sun, fun and rum deal: Perspectives on development in the Commonwealth Caribbean. *The Caribbean* (Winter), 19–26.

Robinson, D.J. (1989) The language and significance of place in Latin America. In J.J. Agnew and J.S. Duncan (eds) *The Power of Place* (pp. 157–184). Boston: Unwin Hyman.

Rohter, L. (1998) A tiger in a sea of pussycats: Trinidad and Tobago bids goodbye to oil, hello to gas. *New York Times*, 4 September.

Sampath, N. (1997) Mas identity: Global and local aspects of Trinidad Carnival. In S. Abram, D. Macleod and J.D. Waldren (eds) *Tourists and Tourism: Identifying with People and Places* (pp. 149–171). New York: Routledge.

Sauer, C.O. (1925) The morphology of landscape. *University of California Publications in Geography* 11 (2), 19–54.

Sautman, F. (1982) The quick and the dead in the communal feasts of Ashura and Carnival. *Comparative Civilizations Review* 9 (Fall), 45–85.

Smart, I.I. and Nehusi, K.S.K. (2000) *Ah Come Back Home: Perspectives on the Trinidad and Tobago Carnival*. Port of Spain: Original World Press.

Stuempfle, S. (1995) *The Steelband Movement: The Forging of a National Art in Trinidad and Tobago*. Philadelphia: University of Pennsylvania Press.

Wilkes, J. (2002) What if Christ came to T & T on carnival day? *Trinidad Express Newspaper*. See www.trinidadexpress.com (accessed January 2002).

Yawching, D. (2002) Carnival musings. *Trinidad Express Newspaper*. See www.trinidadexpress.com (accessed February 2002).

Zelinsky, W. (1992) *The Cultural Geography of the United States*. Hoboken: Prentice Hall.

Part 3
Adjustment and Change

9 Retreating towards Subjective Well-Being

Melanie Kay Smith

Introduction

This chapter approaches well-being from the perspective of Aristotle, who is often considered to be the 'founder' of subjective well-being or happiness. In addition to pleasure-seeking and hedonic experiences, he emphasises the importance of eudaimonia, which includes psychological well-being and the realisation of human potential, as well as meaningful actions (Boniwell, 2008). Whereas hedonic pleasure-seeking activities can provide instant well-being, eudaimonic effects might result from activities that are more challenging at the time, but which afford delayed positive benefits (Knobloch et al., 2016). The quest for 'existential authenticity' can be challenging but forms an integral part of well-being (Kirillova & Lehto, 2015). This idea will be explored further within the context of holistic retreat centres, which can be psychologically or emotionally challenging but also have cathartic effects contributing to personal transformation (Fu et al., 2015; Heintzman, 2013; Lea, 2008; Reisinger, 2013; Smith, 2013). Such centres are often constructed around activities that promote personal growth, self-fulfilment and self-development, some of the cornerstones of eudaimonia (Cloninger, 2004). Retreats ideally provide a combination of pleasure and hedonism (i.e. maximising pleasure and minimising pain), altruistic activities (e.g. being environmentally friendly or benefitting local communities) and meaningful experiences (e.g. education, self-development) (Smith & Diekmann, 2017). Many retreats use Eastern therapies to encourage 'self-transcendence', which is also an important element of well-being (Joshanloo, 2014). The data collection uses discourse analysis of retreat websites, employing a purposive sampling approach, and also includes participant reviews (after Fu et al., 2015). The main aim was to analyse activities and programmes that purport to contribute to subjective well-being via the discourse that is used to promote their benefits and how they are experienced by participants.

Tourism and Well-Being Revisited

This chapter focuses on the subject of tourism and well-being from a tourist's perspective and its role in generating meaningful, eudaimonic and transformational experiences. The original concept of well-being was connected to the human desire to experience as much pleasure as possible and to avoid pain (Carlisle *et al.*, 2009). It has been suggested that the conceptualisation of well-being originates from two different philosophical traditions: the hedonic approach and the eudaimonic approach. The former is associated mainly with happiness, whereas the latter includes self-actualisation and fulfilling one's potential (Ryan & Deci, 2001). It has sometimes been argued that intellectual and aesthetic pleasures have a 'higher value' than bodily pleasures, an idea that stems back to the utilitarian philosophies of both Bentham (1789) and Mill (1863), who argued that some pleasures are more valuable than others (Heathwood, 2014). In the context of tourism, this might suggest that an education-orientated cultural trip has more 'value' than a relaxation-based 'sun–sea–sand' holiday. However, this chapter ideally wants to exclude such value judgements and argue that all pleasures have their place and time in human happiness. After all, if well-being is partly about satisfying idealised desires (Heathwood, 2014), tourism is the perfect activity for that. Fennell (2009) provides a comprehensive analysis of the role pleasure plays in tourism, including anticipation of trips, novelty-seeking, and even transgressive behaviours. It has been argued that happiness is by nature episodic (Feldman, 2008; Raibley, 2011) and tourism affords moments of hedonia within specific time constraints. 'Objective list' theories of well-being (Alexandrova, 2012) emphasise the longer-term impacts of (hedonistic) behaviour and suggest that people do not always know what is good for them at the time (Heathwood, 2014). While not wanting to denigrate hedonistic and pleasurable tourism experiences, the importance of eudaimonia should also be considered.

Aristotle emphasised that pleasure-yielding desires will ultimately not enhance well-being as much as leading a virtuous, meaningful life and fulfilling human potential (Boniwell, 2016). Eudaimonia incorporates the notion of realising human potential more than pleasure-seeking and is connected to personal growth, self-fulfilment and meaningful behaviour (Cloninger, 2004). As suggested by Filep (2012) and Smith and Diekmann (2017) based on Seligman's (2002) authentic happiness model (pleasant life, good life, meaningful life), the ideal well-being-enhancing tourism experience should revolve around pleasure, altruistic activities and meaningful experiences. Eudaimonia may even include elements of hardship to reach a specific goal and the positive effects can therefore be delayed (Knobloch *et al.*, 2016). Examples of this might include extreme adventure, volunteering, pilgrimage or retreats. The latter offer self-development workshops that allow participants to move forward in their lives or even transform but can be psychologically and emotionally painful at the time (Fu *et al.*, 2015; Heintzman, 2013; Lea, 2008; Reisinger, 2013; Smith, 2013). Wong (2016) argues that self-transcendence is also essential for well-being, along with continual self-improvement and full realisation of one's potential. Williams and Harvey (2001) define transcendence as moments of subjective awareness, intense happiness and freedom, as well as a sense of harmony with the entire world. Part of Aristotle's notion of eudaimonia was also connected to universal harmony (Sørensen, 2010).

The types of holidays chosen by tourists may reflect their cultural background, their age or life stage, their current state of needs, or even their personality. Clearly, the timing and choice of a specific destination or activities will affect well-being outcomes differently, even though little is known about how behavioural choices affect tourists' subjective well-being (Mitas *et al.*, 2016). Table 9.1 provides a summary of tourism practices and how they relate to well-being concepts, adapted from Smith and Diekmann (2017).

It is, of course, beyond the scope of this chapter to explore all dimensions of well-being and the concomitant tourism experiences. The focus in this chapter will be mainly on those experiences that offer eudaimonic benefits (without excluding some hedonistic elements). This includes forms of tourism where the tourist is actively seeking some self-development or personal transformation. It could be argued that such forms of tourism ultimately lead to greater happiness or subjective well-being in everyday life and not just in the temporary context of tourism. This form of eudaimonia has been described by Şimşek (2009: 509) as the affective dimension of subjective well-being, which is 'related to personal goals and

Table 9.1 Tourism practices and well-being concepts

Main aims	Hedonic well-being	Hedonic and eudaimonic well-being	Eudaimonic well-being	Utilitarian well-being
Characteristics	To experience as much pleasure and enjoyment as possible	Combinations of pleasurable and relaxing experiences with educational or self-development ones	Self-fulfilment, existential authenticity, transformation, altruism	Creating the greatest good for the greatest number of people within the limits of the earth's resources
Examples	Beach holidays, party and alcohol tourism, wellness spas	Cultural tourism followed by nightlife; adventure tourism followed by beach relaxation	Self-development programmes in holistic retreats; volunteer tourism and charity treks	Responsible ecotourism; community-based Indigenous tourism

Source: Smith & Diekmann (2017).

projects, the destination of which is growth, meaning, and self-realization in a purely personal / phenomenal sense'. The temporal element of eudaimonia is also significant here, as emphasised by Aristotle and existential philosophers like Martin Heidegger. This means that life is assessed not only in terms of goals but also as a 'project of becoming'. The notion of becoming is related to the central theme of 'being-in-the-world' in the work of Heidegger and the human consciousness of what it means 'to be'. To simplify, according to Heidegger, humans do not just exist physically and biologically but are free to choose the kind of life they wish to lead and how they interact with their environment and those around them. In reality, there are constraints and obstacles to 'freedom' (including one's cultural context), but a certain process of becoming can take place within the time span of human life (according to Heidegger Stanford Encyclopedia of Philosophy, 2011).

Eudaimonic Tourism Experiences

Sørensen (2010) summarises the Aristotelian approach to ethics, which is strongly grounded in teleology, by stating that the end or goal (*causa finalis*) of human existence is happiness or flourishing (eudaimonia), achieved through a balanced use of human reason or virtues in relation to everyday life in society as well as the contemplation of universal harmony. Aristotle argued that man is a social being as well as an individual, but whereas politics is more concerned with the flourishing of society, ethics is the art of the good life dealing with the flourishing of the individual. Ancient ethics was an art of living that meant getting the soul into balance whereas the art of medicine was more about getting the body into balance (Sørensen, 2010). The connection between Aristotle's idea of eudaimonia and existential philosophy is quite strong in that it promotes the character of individuals and the development of personal virtues (Davenport, 2001).

Existentialism as a philosophy can provide some interesting insights into the process of human becoming (per Heidegger), which is arguably connected to eudaimonic practices. Existentialism is concerned primarily with the problematic nature of individual human existence or being in the world. Its proponents argue that although humans are technically free to choose the nature of their existence, they are constrained historically and contextually. Nevertheless, the theory of Heidegger and Jean-Paul Sartre that 'existence precedes essence' is perhaps the most relevant here, meaning that humans do not have a predetermined nature but can create and project themselves into the world (Abbagnano, 2021). Existentialist philosophers like Nietzsche and Kierkegaard praised the idea of living for oneself and creating and determining one's own virtues. Kierkegaard also emphasised the importance of living authentically, which has a strong connection to ethics. Sartre's notion of (ontological) freedom allows individuals to become their future selves and to adopt new projects or values at any time (even though 'practical' freedom may limit this).

Tourism could perhaps be described as a process of losing and finding oneself (Iyer, 2000). However, this notion presupposes that there is a fixed self to be found. Gazley and Watling (2015) argue instead that the travelling self is created using symbolic products and experiences. It is composed of a combination of 'selves' that are partly ideal and social – that is, connected to how one would like to be seen by others (Sirgy & Su, 2000). Nevertheless, transformational tourism literature has suggested that there is a true or authentic self that can emerge or be revealed through travel (Brown, 2013; Reisinger, 2013; Smith, 2013).

Existentialist studies highlight the connection of tourism to anxiety, dread or death (Brown, 2013; Kirillova & Lehto, 2015). The existentialist writer Albert Camus (1962) suggested that travel was an occasion for testing the spirit and not only for pleasure. The hedonic aspects of tourism might only 'tranquilize' existential anxiety in the short term (Kirillova & Lehto, 2015) as a form of release or 'temporary forgetting' (de Botton, 2009). More eudaimonic forms of tourism might, however, result in longer-term benefits as individuals move closer to a sense of existential authenticity, self-realisation, or even transformation (Reisinger, 2013). It will be argued in this chapter that retreat holidays offer an ideal opportunity to combine relaxation with escapism, engagement with the 'authentic' self and the potential for transformation.

Motivations for and Benefits of Retreats

The concept of 'retreat' may mean a physical place or an opportunity or moment in time for rest, reflection, or self-improvement of some sort (Kelly & Smith, 2017). Retreats offer a space in which to relax and replenish personal health and development and are no longer associated only with spiritual or religious needs (Glouberman & Cloutier, 2017). Some retreats also offer health- and lifestyle-enhancing complementary therapies promoting, for example, weight loss or drug and alcohol rehabilitation (Cohen *et al.*, 2017). Typical activities include yoga, meditation, fitness, counselling, stress management, healthy nutrition and detoxification, creative practices, reconnecting to nature, and, more latterly, digital detox and post-COVID recovery. The length of time spent in a retreat can vary from one day to a long weekend or, most commonly, a week. Norman and Pokorny (2017) note that meditation retreats can range from an overnight stay to a year in length, but the most common length is around 6–10 days.

In terms of motivation for visiting retreats, Kelly's (2012) research shows that the main motivations for going to retreats are to destress and unwind, as well as enhancing health more generally. She noted that the rest/relaxation factor was the main motivator, followed to a lesser extent by social and spiritual reasons. Kelly and Smith (2017) suggested that retreat-based wellness tourism can help people to discover their 'true' or 'authentic' selves (a eudaimonic process). This can happen through the focus on different domains of one's life and balancing or harmonising those domains. However, it should be noted that retreat centre activities may also reveal aspects of one's character that are not positive or are obstructive to self-development. The rhetoric of retreats implies that participants should work through such deficiencies to emerge a 'better person'.

The majority of retreats seem to include yoga as a central activity. Yoga tourism has been identified as a transformative type of tourism that can lead to self-actualisation and spiritual renewal (Ponder & Holladay, 2013) and help people return to their authentic selves, as well as enhancing their emotional and spiritual well-being (Reisinger, 2013). Although the origins of yoga are deeply spiritual, many modern-day practitioners may prefer to think that they are 'working on themselves' (Garrett, 2001) and can gain spiritual or non-spiritual benefits depending on their personal and individual preferences (Bowers & Cheer, 2017). Indeed, Heelas and Woodward (2005) identified yoga as a practice that helps to facilitate the convergence of the personal and spiritual paths. However, even if novice yogis firstly experience the practice as mainly physical, mental and spiritual transformation can happen with regular practice (Smith & Sziva, 2017).

Spiritual activities within retreats are still somewhat integral even if they are not always advertised explicitly. Bone's (2013) research suggests that the spiritual experience of being in a retreat is pervasive, regardless of whether participants gain the most benefits from escapism, forming part of a community, or being in a therapeutic setting. Other authors have noted that tourists go to retreats with a non-religious motivation but still experience the sacred or spiritual. They divide such 'sacred-spiritual' experiences into the 'Search for Meaning' and the 'Search for Escape': the former refers to the focus on one's inner being, whereas the latter refers to connections and relationships with the world and others (Jiang *et al.*, 2018). Spiritual tourism may also include a search for personal meaning, elements of transcendence and connectedness with oneself and others (Wilson *et al.*, 2013), as well as personal challenges and adventure, which lead to greater fulfilment (Cheer *et al.*, 2017).

In addition to a range of activities and therapies, the natural environment or landscape around the retreat can be important to the experience (Ashton, 2018; Jiang *et al.*, 2018). For example, research has shown that quiet and remote rural areas like mountains tend to offer the most transcendental experiences (Sharpley & Jepson, 2011), while deserts offer a deeply spiritual experience (Moufakkir & Selmi, 2018). The health-enhancing effects of blue spaces are also well documented (Gascon *et al.*, 2017).

The experience of visiting a retreat is an intense one and, although greater happiness may be derived in the long term, the tourist may pass through some uncomfortable moments in connecting to their 'true self' (Lea, 2008). Many tourists who visit retreats have recently experienced negative life events such as serious illnesses, relationship break-ups, or the death of a loved one (Heintzman, 2013), or may be facing existential, physical, or work-related challenges (Fu *et al.*, 2015). Lea (2008) emphasises the benefits of removing oneself from everyday life in order to rest and recuperate; however, escapism is only one part of the experience in retreat tourism. Benefits can include improved mood, greater control over life and making positive adjustments to lifestyle that lead to health improvements (Cohen *et al.*, 2017). Self-development and longer-term transformation are also a major focus of this type of tourism (Smith, 2013). Retreats may also encourage participants to rediscover the joy in their lives, an approach advocated by Dina Glouberman (2003), who co-founded the still successful Skyros holistic holidays in 1979. She also emphasises the importance of being with like-minded communities (Glouberman & Cloutier, 2017). Literally thousands of retreat centres now exist worldwide, so it is interesting to delve deeper into exactly what is offered and how it can benefit tourists' subjective well-being in different ways longer-term.

A Discourse Analysis of Retreat Experiences

The following section provides an analysis of the retreat sector based on website discourse analysis. Like most netnographic studies in tourism, this study is textual rather than visual, using thematic analysis and focused on static text (Tavakoli & Wijesinghe, 2019). Although the role of the researcher was 'passive' and non-participatory on this occasion, involvement with the retreat sector has been ongoing for 15 years through research, participation, and acquaintance with retreat organisers. In accordance with Kozinets *et al.* (2014) the researcher concurs that prolonged engagement and immersion in online communities is necessary to gain deep understanding and more reflexive retreat studies should also include participant observation, which COVID-19 unfortunately precluded.

The sampling was purposive and based on the author's selection (and frequent use) of a retreat portal that has existed for 25 years, and which contains a large and diverse range of retreats. Special attention was paid to the way in which the retreats are categorised in terms of typologies, the activities that are contained therein, and the purported benefits. The focus was on the qualitative aspects of language rather than taking a systematic or quantitative approach, thus differentiating it from content analysis (Franzosi, 2008; Shaw & Bailey, 2009). Clearly, all website promotion includes persuasive linguistic strategies (Malenkina & Ivanov, 2018), including metadiscourse (Hyland & Tse, 2004), which can help to identify the interpersonal, social, and cultural resources writers use to present propositions and persuade or engage readers. The language of retreats tends to be rather personal and emotive, pertaining to tourists' deepest feelings, needs, desired outcomes and benefits, and even fears. In accordance with the theories of Hannam and Knox (2005), such discourses could be described as mediated cultural products that are part of wider systems of knowledge. In the case of retreat holidays, the sociocultural context is mainly that of relatively affluent Western tourists who tend to be 'money rich, time poor' in professional jobs. Their tendency towards overcommitment and optimal performance can easily result in stress and burnout, as documented in Glouberman (2003).

Discourse analysis of the website of The Retreat Company

The sampling in this data collection was based on the portal of The Retreat Company (www.theretreatcompany.com), which features hundreds of retreats in 17 or more countries across six continents. This portal was chosen because The Retreat Company is one of the longest running retreat portals, available since 1996. It is difficult to estimate the number of retreats or retreat holidays in the portal currently as they feature multiple times under different topic headings, but an estimate would be 500 different types of retreat holiday. The website features various search options include Location (i.e. country) and Topics. Under the category of All Topics, there are currently around 100, ranging from Addictions to Fitness to Raw Food to Yoga. Table 9.2 shows the author's categorisation of these topics, excluding those that include types of accommodation (e.g.

Table 9.2 A categorisation of topics featured in The Retreat Company's portal

Category of retreat	Topics
Physical health/bodywork	Addiction, Bodywrap, Bootcamp, Cancer Patients, Colon Cleansing, Massage, Pilates, Reflexology, Stop Smoking
Fitness and sports	Adventure and Sports, Cycling, Fitness, Horse Riding, Surfing, Walking and Hiking
Nutrition	Cookery, Nutrition, Organic, Detox, Juice, Raw Food
Mental/emotional/psychological	Anxiety, Bereavement, Cognitive Behavioural Therapy (CBT), Counselling, Depression, Emotional Freedom Technique, Hypnotherapy, Mindfulness, Neuro-linguistic Programming, Psychotherapy
Life management/self-development/stress management	Life Coaching, Burnout, Corporate Career Retreat, Digital Detox, Divination, Divorce Recovery, Executive, Fertility, Grow, Holistic Health, Menopause, Natural Health, Personal Development, Recovery, Recuperation, Relationships, Stress Management, Weight Loss and Management
Yoga/meditation	Breathwork, Meditation, Silence, Sound Healing, Yoga
Creative	Arts & Creativity, Dance, Music, Singing, Writing
Eastern/spiritual	Ayurveda, Buddhist, Kirtan (sacred chant), Panchakarma, Qi Gong, Reiki, Sacred Journeys, Shamanic, Tai Chi, Tantra, Taoism
Nature/location-based	Coastal, Eco, Nature, Outdoor, Wildlife Experience
Segment-based	Couples, Family, Men Only, Women Only

cruise, bed and breakfast) or those that feature retreats to buy or rent. Manual inductive coding was used to group the retreats into categories according to the focus and activities offered.

It is important to note that The Retreat Company is based in England, UK, and only features those retreats that run programmes in English. Although the portal can be accessed by anyone in the world, this already provides useful information about the sociocultural context for the discourse analysis. Retreat holidays usually attract tourists who are middle-aged on average; solo travellers are encouraged and some retreats do not even allow children. For example, Kelly's (2012) research on retreats showed that the age range varied but a relatively high number of retreat goers were between 35 and 55 and around 88% of them were women. Skyros (established in 1979) estimates that around 70% of their guests are aged 35–65 and that over 70% are solo travellers. Discourse analysis should therefore take into consideration that retreat centres are largely female-dominated spaces, attracting solo travellers without children who are likely to be middle-class, educated professionals (Kelly & Smith, 2017).

The Retreat Company portal emphasises health, rest, recuperation, escapism, but also transformation in its opening promotion, 'When you are looking for a healthy holiday, a rest, somewhere to recuperate, a bolt hole or transformation vacations, our online retreats directory delivers a range of retreats and topics to choose from'. The word 'bolt hole' has become a popular term in British journalese and blogs, meaning a place where a person can escape and hide from feeling busy and overwhelmed. The Retreat Company describes itself as an 'authentic business', which also taps into the growing importance of authentic experiences in tourism (including existential authenticity). The stated purpose of the company refers to several of the issues that have been mentioned so far in the chapter, including disconnection from one's true self:

> We have become out of touch with our true essence, of who we really are. This is partly because of the culture of speed, stress and over-reliance on technology. In the world where we have vast evolution of technology that is upgraded moment by moment. Plus the demands made on us, by ourselves, and the outside world 'not enough hours in the day' concepts are spiralling out of control. Stress, crisis management and burnout are commonplace. (The Retreat Company, 2022)

It is suggested that personal balance helps to deal with world unrest and that world peace starts with inner peace. This connects back to Aristotle's notion that eudaimonia is based on both personal flourishing and universal harmony.

It was planned that systematic sampling would be used to analyse the retreat centres or holidays offered by selecting every fifth retreat holiday. However, the categorisation of holidays by location and topic, as well as the wider categorisations of Health & Wellness or Yoga, followed by the various spotlights (e.g. on Detox or Spiritual), complicated this sampling strategy. Thus, purposive sampling was used instead to select one example of a retreat from each category identified by the author in Table 9.2 (e.g. physical health/bodywork, fitness and sports). Retreats were selected whose programmes and activities corresponded most closely to eudaimonic outcomes or benefits. To reiterate, according to Aristotle and subsequent theorists' definitions of eudaimonia, this includes psychological well-being, getting the soul in balance, development of personal character and values, living an ethical or virtuous life, living authentically, meaningful actions, personal growth, fulfilling human potential, life as a project of becoming, and contributing to universal harmony. This generated text from 10 retreats, although numerous retreat programmes were explored before the final selection was made. Table 9.3 presents some examples of retreats from the different categories and connects the activities to the dimensions of eudaimonia previously listed.

The large number of active verbs should be noted in the discourse, which relate to eudaimonic personal growth and development – for example, 'to improve', 'to enhance', 'to cultivate', 'to learn', 'to become'. The participant is encouraged to improve themselves and their lives with the use of verbs like 'to heal', 'to grow', 'to reconnect', 'to cleanse', 'to rebalance', 'to renew' and to become more adventurous, joyful or energetic. These actions are accompanied by a promise of support from the retreat organisers – for example, 'to encourage', 'to enable', 'to teach', 'to help', 'to coach', 'to guide'. The tone of the promotional discourse is nurturing, reassuring, and kind. The use of collective pronouns like 'we' and 'us' give the impression that the individual is not alone for the duration of the retreat and is engaged in a process of cocreation as well as being part of a community. Ultimately, the discourse revolves around a process of supported personal development that allows the participant to return to a life of 'becoming' and which involves personal growth and betterment. This is reflected in words and phrases like 'to heal and grow', 'for life ahead', 'the person you were meant to be', 'your purpose', 'a clear sense of your priorities', 'the future well-being journey of You', 'a new path' and 'lasting change'.

Visitor comments and testimonials were also selected from the most featured retreats that offer permanent, accessible websites and were not one-offs or run by an individual. At the time, there were relatively few because of COVID-19, so these comments were mainly taken from 2019. A large number of comments refer to the experience as joyful, life-enhancing and intensely memorable – for instance, 'The joy of living is much stronger now after the weekend together'. Here, the use of the collective adverb 'together' should be noted, indicating a feeling of support and being with others. Several comments refer to the benefits of escapism from stressful everyday life: 'It was excellent for me to retreat from "civilisation". . . . It has been awakening, energising, restful, interesting and freeing'. The verbs echo those in the promotional discourse implying a

Table 9.3 Discourse analysis of selected retreats

Topic	Category (see Table 9.2)	Activities	Discourse examples	Connection to eudaimonia
Anxiety and Depression Recovery	Mental/emotional/ psychological	Use of CBT, positive psychology, yoga, massage, mindfulness, and mindful self-compassion	Feel 'emotionally balanced and joyful with excitement for life ahead' 'Become the person you were meant to be'	Getting the soul in balance Fulfilling human potential
Wellness and Wellbeing retreats	Life management / self-development/ stress management	Life coaching sessions, life balance assessment, personal behaviour mapping, stress relief program, mindfulness sessions, family constellation, nutritional advice	'We offer real support, coaching and guidance for the future well-being journey of You' There is talk of 'a new path', 'your purpose in the world', with a longer-term 'concrete plan of action for lasting change'	The project of 'becoming' Virtuous and meaningful actions Contribution to universal harmony
Restore and Revitalise Retreats – for Cancer 'Thrivors' and 'Survivors'	Physical health/ bodywork	Daily yoga and meditation classes, complementary therapies, talks on nutrition, plenty of time to rest and relax	'You will leave feeling relaxed, in better health, and strengthened so that you can continue the treatment process with renewed energy and vitality, step back into your normal life with a feeling of wholeness and direction, and have a clear sense of your priorities'	Psychological well-being Getting the soul in balance Fulfilling human potential
Transformational Spiritual Retreats for Living and Vibrant Harmony	Eastern/spiritual	Primordial sound meditation, vibrant sounds of nature	'Get away from it all' 'Reconnect with your essence'	Psychological well-being Connection to the authentic or true self

(Continued)

Table 9.3 Discourse analysis of selected retreats

Topic	Category (see Table 9.2)	Activities	Discourse examples	Connection to eudaimonia
Wellbeing for Mind & Body	Yoga/meditation	Yoga, meditation, walks in nature	Yoga is described as a practice that 'connects us to the limitless nature of the true self'. Participants are taught how to reconnect with nature and to 'cultivate more joy in your connection with others'	Authentic and meaningful living Development of personal character and values Contribution to universal harmony (nature and social)
Nutrition retreats	Nutrition	Cookery, nutrition, organic food consumption, detox, juicing, raw food	'This is a wonderful program, which will help to cleanse, detoxify and rebalance the body' 'Nutritional therapy can bring counselling and therapy to enable healthy attitudes to food and diet'	Psychological well-being Development of personal character and values
Circle dancing	Creative	Dance, music, singing	'We create the magic of dancing together, moving as one, yet distinct as individuals'	Contribution to universal harmony (social) Psychological well-being
Fitness & Wellness retreats	Fitness and sports	Yoga, Pilates, fitness classes, hiking, cycling, windsurfing, horse riding	'A wonderful retreat encouraging you to live a healthy and adventurous lifestyle' 'You are thoroughly encouraged to get out and about in an eco-friendly way'	Personal growth Fulfilling human potential Contribution to universal harmony (nature)
Unwind and enjoy in nature	Nature/location-based	Coastal, eco, nature, outdoor, wildlife experience	'We'll spend as much time as we can outside, reminding ourselves how therapeutic nature and outdoors can be by trying out some "ecotherapy"'	Psychological well-being Getting the soul in balance
Couples retreat	Segment-based	Meditation, juice detox, and relationship coaching	'To improve communication, enhance intimacy and learn tools to heal and grow'	Development of personal character and values Personal growth

sense of growth and change (i.e. being 'awakened' and 'energised') as well as gaining (existential) freedom from the shackles of everyday life.

References are made to (re)gaining a sense of the true self, authenticity and transformation: 'This is a place of gentle, natural transformations, a space away from the madness of everyday life where I find peace, inspiration, hope, love and most importantly a sense of who we are'. Again, the use of the collective pronoun 'we' should be noted as it implies a process that is collective and not only individual. The reference to 'love' also indicates support and 'gentle' suggests nurturing. Both comments note the need to 'retreat' or escape from everyday life. A further comment also emphasises the need to find peace with the use of the word 'stillness'. 'Gratitude' seems to pay tribute to the retreat organisers who have helped the participant: 'My deepest gratitude for holding space where I could be authentic, dig deep and find stillness and radiance.' Use of the word 'space' was common in the participants' discourse, which would not necessarily be found in everyday life (hence the need to travel to retreat). The word 'authentic' recalls ancient ethics and existential theory in the process of personal growth and discovery of the 'true' essence or self. Several participants refer to change, self-development and new beginnings, often using the image of 'a journey': 'I've taken away lots from the weekend that I know has added to my journey in life' and 'it has been a life changing and totally spiritual journey for me'. This is part of the life project of 'becoming' as propounded by Aristotle and Heidegger.

People also refer to the fact that the retreat encouraged them to become their best selves and to flourish: '[T]he retreat changed my outlook and encouraged me to follow my dreams'. The natural environment frequently enhanced the stay, suggesting that reconnecting to nature can be essential to the notion of 'universal harmony': 'I learnt how to stop battling with myself and understood the healing powers of nature' and 'a place I love to visit to reconnect with nature and touch peace'. Once again, the need for 'peace' is mentioned as well as the process of 'healing' and struggling with the self to overcome those issues that impinge on personal growth and freedom.

Conclusion

Retreat centre discourse contains frequent references to realising human potential, personal growth, self-development and self-fulfilment in accordance with theories of eudaimonia (Boniwell, 2008; Cloninger, 2004). Similar vocabulary seems to be employed in both promotional and participant discourse with a strong theme of the supportive and collective nature of the retreats. It is implied that such processes could not take place alone at home and that a retreat 'space' is needed to escape, rest, heal and grow. In addition to self-development, learning new skills is also encouraged, whether they are creative or about self-care and good health. These processes form part of 'meaningful experiences' as indicated in Seligman's (2002) notion of authentic happiness. Altruism is fostered indirectly by encouraging participants to reconnect with nature and to form communities that support each other. This also connects closely to Aristotle's idea of 'universal harmony', which should include both nature and society. In addition to the supportive facilitators and community in retreats, the natural landscape is presented as an important element in the relaxation and healing process, corroborating suggestions by Ashton (2018) and Jiang *et al.* (2018).

Stress management is still a major motivation for going to retreats, as indicated by Kelly (2012), along with rest and relaxation. Numerous references are made to 'rejuvenating', 're-energising', 'recharging', 'replenishing', 'rebooting', and so on. Hedonic pleasure is not excluded either, as there is much discussion of 'passion' and 'joy' too. Nevertheless, it is indicated that participants in retreats may need to overcome some of the obstacles or barriers that are holding them back from living their best lives. This might represent the 'practical freedom' identified by Sartre, which contains obstacles and barriers compared to 'ontological freedom'. This could require some more challenging workshops that afford delayed benefits (Heintzman, 2013; Knobloch *et al.*, 2016; Lea, 2008) and such practices perhaps come closer to encountering the darker side of life – but still in the context of well-being (Nawijn & Filep, 2016). Words like 'spirituality' and 'transcendence' are used quite sparingly in retreat discourse, but spiritual practices are pervasive, as suggested by Bone (2013). It is clear that most retreat centres and their participants believe wholeheartedly that there is a 'true self' to be (re)discovered, developed and transformed, which confirms previous studies (Brown, 2013; Kelly & Smith, 2017; Reisinger, 2013; Smith, 2013). Reference is also made to being or becoming 'authentic', which connects to Kierkegaard's emphasis on the need for 'authentic living' or Heidegger's notion of life as a project of 'becoming'.

Overall, it seems that those retreats that focus on the affective dimensions of well-being afford the most eudaimonic experiences or benefits; however, bodywork, fitness, nutrition, creativity and spiritual practices can all assist in the process of becoming. Retreat discourse makes frequent reference to life as a 'journey' or 'path' and how retreats' carefully combined programmes and activities can help participants to get back on track or to explore new directions. It could be concluded that retreats are life-enhancing in myriad ways, affording more than just episodic happiness and hedonic well-being. The changes engendered appear to be deeper, longer lasting and often truly eudaimonic.

References

Abbagnano, N. (2020) Existentialism. Encyclopedia Britannica. See www.britannica.com/topic/existentialism (accessed July 2021).

Alexandrova, A. (2012) Well-being as an object of science. *Philosophy of Science* 79 (5), 678–689.

Ashton, A.S. (2018) Spiritual retreat tourism development in the Asia Pacific region: Investigating the impact of tourist satisfaction and intention to revisit: A Chiang Mai, Thailand case study. *Asia Pacific Journal of Tourism Research* 23 (11), 1098–1114.
Bentham, J. (1789) *An Introduction to the Principles of Morals and Legislation.* Oxford: Clarendon Press.
Bone, K. (2013) Spiritual retreat tourism in New Zealand. *Tourism Recreation Research* 38 (3), 295–309.
Boniwell, I. (2008) *Positive Psychology in a Nutshell.* London: Personal Well-Being Centre.
Boniwell, I. (2016) The concept of eudaimonic wellbeing. Positive Psychology UK. See www.positivepsychology.org.uk/pp-theory/eudaimonia/34-the-concept-of-eudaimonic-well-being.html (accessed June 2021).
Bowers, H. and Cheer, J.M. (2017) Yoga tourism: Commodification and western embracement of eastern spiritual practice. *Tourism Management Perspectives* 24, 208–216.
Brown, L. (2013) Tourism: A catalyst for existential authenticity. *Annals of Tourism Research* 40 (1), 176–190.
Camus, A. (1962) *Carnets: Mai 1935 - février 1942.* Paris: Gallimard
Carlisle S., Henderson, G. and Hanlon, P.W. (2009) Wellbeing: A collateral casualty of modernity? *Social Science and Medicine* 69, 1556–1560.
Cheer, J.M., Belhassen, Y. and Kujawa, J. (2017) The search for spirituality in tourism: Toward a conceptual framework for spiritual tourism. *Tourism Management Perspectives* 24, 252–256.
Cloninger, R.C. (2004) *Feeling Good: The Science of Well-Being.* Oxford: Oxford University Press.
Cohen, M.M., Elliott, F., Oates, L., Schembri, A. and Mantri, N. (2017) Do wellness tourists get well? An observational study of multiple dimensions of health and wellbeing after a week-long retreat. *The Journal of Alternative and Complementary Medicine* 23 (2), 140–148.
Davenport, J.J. (2001) Towards an existential virtue ethics: Kierkegaard and MacIntyre. In J.J. Davenport and A. Rudd (eds) *Kierkegaard after MacIntyre* (pp. 265–324). Open Court Publishing Co.
De Botton, A. (2009) *A Week at the Airport.* London: Profile Books.
Feldman, F. (2008) Whole life satisfaction concepts of happiness. *Therria* 74, 219–238.
Fennell, D.A. (2009) The nature of pleasure in pleasure travel. *Tourism Recreation Research* 34 (2), 123–134.
Filep, S. (2012) Moving beyond subjective well-being: A tourism critique. *Journal of Hospitality and Tourism Research* 38 (2), 266–274.
Franzosi, R. (2008) Content analysis: Objective, systematic, and quantitative description of content. In R. Franzosi (ed.) *SAGE Benchmarks in Social Research Methods: Content Analysis* (pp. 2–43). London: SAGE.
Fu, X., Tanyatanaboon, M. and Lehto, X.Y. (2015) Conceptualizing transformative guest experience at retreat centres. *International Journal of Hospitality Management* 49, 83–92.
Garrett, C. (2001) Transcendental meditation, reiki and yoga: Suffering, ritual and self-transformation. *Journal of Contemporary Religion* 16 (3), 329–342.
Gascon, M., Zijlema, W., Vert, C., White, M.P. and Nieuwenhuijsen, M.J. (2017) Outdoor blue spaces, human health and well-being: A systematic review of quantitative studies. *International Journal of Hygiene and Environmental Health* 220 (8), 1207–1221.
Gazley, A. and Watling, L. (2015) Me, my tourist self and I: The symbolic consumption of travel. *Journal of Travel & Tourism Marketing* 32, 639–655.
Glouberman, D. (2003) *The Joy of Burnout: How the End of the World Can Be a New Beginning.* London: Hodder & Stoughton.

Glouberman, D. and Cloutier, J. (2017) Community as holistic healer on health holiday retreats: The case of Skyros. In M.K. Smith and L. Puczkó (eds) *The Routledge Handbook of Health Tourism* (pp. 152–167). London: Routledge.

Hannam, K. and Knox, D. (2005) Discourse analysis in tourism research: A critical perspective. *Tourism Recreation Research* 30 (2), 23–30.

Heathwood, C. (2014) Subjective theories of well-being. In B. Eggleston and D.E. Miller (eds) *The Cambridge Companion to Utilitarianism* (pp. 199–219). Cambridge: Cambridge University Press.

Heelas, P. and Woodhead, L. (2005) *The Spiritual Revolution: Why Religion is Giving Way to Spirituality*. Malden, MA: Blackwell Publishing.

Heintzman, P. (2013) Retreat tourism as a form of transformational tourism. In Y. Reisinger (ed.) *Transformational Tourism: Tourist Perspectives* (pp. 68–81). Oxford: CABI.

Hyland, K. and Tse, P. (2004) Metadiscourse in academic writing: A reappraisal. *Applied Linguistics* 25 (2), 156–177.

Iyer, P. (2000) Why we travel. *Salon*, 18 March. See www.salon.com/2000/03/18/why (accessed June 2021).

Jiang, T., Ryan, C. and Zhang, C. (2018) The spiritual or secular tourist? The experience of Zen meditation in Chinese temples. *Tourism Management* 65, 187–199.

Joshanloo, M. (2014) Eastern conceptualizations of happiness: Fundamental differences with Western views. *Journal of Happiness Studies* 15, 475–493.

Kelly, C. (2012) Wellness tourism: Retreat visitor motivations and experience. *Tourism Recreation Research* 37 (3), 205–213.

Kelly, C. and Smith, M.K. (2017) Journeys of the self: The need to retreat. In M.K. Smith and L. Puczkó (eds) *The Routledge Handbook of Health Tourism* (pp. 138–151). London: Routledge.

Kirillova, K. and Lehto, X. (2015) An existential conceptualisation of the vacation cycle. *Annals of Tourism Research* 55 (1), 110–123.

Knobloch, U., Roberston, K. and Aitken, R. (2016) Experience, emotion and eudaimonia: A consideration of tourist experiences and well-being. *Journal of Travel Research*. 56 (5) 651–662. https://doi.org/10.1177/0047287516650937.

Kozinets, R.V., Dolbec, P. and Earley, A. (2014) Netnographic analysis: Understanding culture through social media data. In U. Flick (ed.) *The SAGE Handbook of Qualitative Data Analysis* (pp. 262–276). London: SAGE.

Lea, J. (2008) Retreating to nature: Rethinking 'therapeutic landscapes'. *Area* 40 (1), 90–98.

Malenkina, N. and Ivanov, S. (2018) A linguistic analysis of the official tourism websites of the seventeen Spanish autonomous communities. *Journal of Destination Marketing & Management* 9 (12), 204–233. https://doi.org/10.1016/j.jdmm.2018.01.007.

Mill, J.S. (1863) *Utilitarianism*. London: Parker, Son and Bourn.

Mitas, O., Nawijn, J. and Jongsma, B. (2016) Between tourists: Tourism and happiness. In M.K. Smith and L. Puczkó (eds) *Routledge Handbook of Health Tourism* (pp. 47–64). Abingdon: Routledge.

Moufakkir, O. and Selmi, N. (2018) Examining the spirituality of spiritual tourists: A Sahara Desert experience. *Annals of Tourism Research* 70, 108–119.

Nawijn, J. and Filep, S. (2016) Two directions for future tourism well-being research. *Annals of Tourism Research* 61, 221–223.

Norman, A. and Pokorny, J.J. (2017) Meditation retreats: Spiritual tourism and well-being interventions. *Tourism Management Perspectives* 24, 201–207.

Ponder, L.M. and Holladay, P.J. (2013) The transformative power of yoga tourism. In Y. Reisinger (ed.) *Transformational Tourism: Tourist Perspectives* (pp. 98–108). Wallingford: CABI.

Raibley, J.R. (2011) Happiness is not well-being. *Journal of Happiness* 13 (6), 1105–1129.

Reisinger, Y. (2013) *Transformational Tourism: Tourist Perspectives*. Wallingford: CABI.

The Retreat Company (2022) About us. See www.theretreatcompany.com/about-us (accessed 26 July 2021).

Ryan, R.M. and Deci, E.L. (2001) On happiness and human potentials: A review of research on hedonic and eudaimonic well-being. *Annual Review Psychology* 52, 141–166.

Seligman, M.E.P. (2002) *Authentic Happiness*. New York: Free Press.

Sharpley, R. and Jepson, D. (2011) Rural tourism: A spiritual experience? *Annals of Tourism Research* 38 (1), 52–71. https://doi.org/10.1016/j.annals.2010.05.002.

Shaw, S.E. and Bailey, J. (2009) Discourse analysis: What is it and why is it relevant to family practice? *Family Practice* 26 (5), 413–419.

Şimşek, O.F. (2009) Happiness revisited: Ontological well-being as a theory-based construct of subjective well-being. *Journal of Happiness Studies* 10, 505–522. https://doi.org/10.1007/s10902-008-9105-6.

Sirgy, M.J. and Su, C. (2000) Destination image, self-congruity, and travel behavior: Toward an integrative model. *Journal of Travel Research* 38 (4), 340–352.

Smith, M.K. (2013) Wellness tourism and its transformational practices. In Y. Reisinger (ed.) *Transformational Tourism: Tourist Perspectives* (pp. 55–67). Wallingford: CABI.

Smith, M.K. and Diekmann, A. (2017) Tourism and wellbeing. *Annals of Tourism Research* 66, 1–13.

Smith, M.K. and Sziva, I. (2017) Yoga, transformation and tourism. In M.K. Smith and L. Puczkó (eds) *The Routledge Handbook of Health Tourism* (pp. 168–180). Abingdon: Routledge.

Sørensen, A.D. (2010) *Philosophy and Therapy of Existence: Perspectives in Existential Analysis*. Aarhus: The State and University Library in Aarhus.

Stanford Encyclopedia of Philosophy (2011) Martin Heidegger. *Stanford Encyclopedia of Philosophy*. See https://plato.stanford.edu/entries/heidegger (accessed July 2021).

Tavakoli, R. and Wijesinghe, S.N.R. (2019) The evolution of the web and netnography in tourism: A systematic review. *Tourism Management Perspectives* 29, 48–55. https://doi.org/10.1016/j.tmp.2018.10.008.

Williams, K. and Harvey, D. (2001) Transcendent experience in forest environments. *Journal of Environmental Psychology* 21, 249–260.

Wilson, G.B., McIntosh, A.J. and Wilson, A.L.Z. (2013) Tourism and spirituality: A phenomenological analysis. *Annals of Tourism Research* 42, 150–168.

Wong, P.T.P. (2016) Meaning-seeking, self-transcendence and well-being. In A. Batthyany (ed.) *Logotherapy and Existential Analysis: Proceedings of the Viktor Frankl Institute* (pp. 311–322). Cham: Springer.

10 Enrichment and Enlightenment from Engagement with History through Heritage-Based Tourism

Michael Fagence

> Although tourists may not experience intense feelings of pleasure in such contexts, this does not mean that such experiences are devoid of positive emotions.... [T]ourists in these settings expect to experience emotions such as compassion, gratitude, interest and awe.
>
> Nawijn & Filep (2016: 222)

Introduction

This chapter is centred on an autoethnographic study of a tourist site – Stringybark Creek in Victoria, Australia – where, on an October day in 1878, the Australian bushranger-cum-outlaw Edward (Ned) Kelly and three of his associates were engaged in a gunfight with a four-man police patrol.

That would seem to be so unlikely a subject for considering tourism as a pathway to hope and happiness that an explanation might be helpful in these introductory paragraphs. Clearly, the event mentioned would not be likely to be one that would be celebrated widely, to be at the core of a pleasure-seeking experience, nor to feature prominently in the advertising and promotional materials of travel agencies; and yet, tourist exposure to the site of a historic event of that kind has the potential to be positive, to contribute to an improved level of knowledge and understanding of what the event was about and even why it happened, and to lead – perhaps – to an improved appreciation of opportunities that might flow from the event in terms of personal character enhancement. This is captured in the quotation from Nawijn and Filep (2016) that heads this chapter, which refers specifically to visits to 'dark' tourism sites. Some of these thoughts might be situated towards one extremity in the idiom of the long bow; so, it has become the purpose of this chapter to a least put forth the proposition that there can be benefits derived from giving time to and expending cognitive and emotional energy on giving consideration to controversial issues from history and from dark tourism sites (Hartmann, 2014; Stone, 2006), learning from the exposure to them through tourism. As another idiom tells us, the proof of the pudding is in the eating, so bon appétit.

It is very much safer to visit the site of the Stringybark Creek historical episode today. There are no bullets flying, no armed protagonists, no horses rushing around uncontrollably; all is calm and quiet. So any visits to the site by tourists in the modern era have to draw on any prior knowledge they have of the event, to engage their imagination and, almost inevitably, to align themselves with the cause of one or the other of the protagonist groups. This is the experiential process at work, becoming enlightened about the circumstances, becoming enriched by appreciating those circumstances, and having an experience that is value-added because of the thoughtfulness that has been given to it. Landsberg's (2004, 2015, 2018) research is helpful here with her conceptualisation of what she has referred to as 'prosthetic memories', whereby people can draw on past experiences or knowledge of past events to imagine 'being there' and 'experiencing the happenings' without leaving the comfort of a home base. It is 'being there' without 'being there'; it is being able to experience a happening without the deprivations (or, in some case, elations) of being there; it is having the time and metaphorical space to indulge in speculations, 'what if' scenarios, and so on without being committed to engagement; it is a pathway to enrichment and enlightenment under circumstances that are physically comfortable, even if they are not psychologically comfortable.

This chapter is heavily predicated on experiencing images and drawing on memories and then converting those images and memories into a new understanding, a new attitude and behaviour, and a new set of values and appreciation of and for the story being told. It is a position that is somewhat consistent with, for example, Seligman's (2002) conceptualisations of 'authentic happiness', where there is a fusion of the good life with the meaningful life – the eudaimonic tradition, a matter that is taken up in some of the case studies in Filep and Pearce (2014) and in Sirgy and Uysal's (2016) recent commentary about a eudaimonia research agenda for travel and tourism.

One of the important tasks for this chapter has been to isolate the very symbols that imbue the selected story with its particular character and to identify which among those are most likely to hold attention, to have the boldest images, to be retained in the memory, and to have the capacity to influence the process of taking stock of what happened. This task has been tackled by considering three pathways that engage in this through tourism; these – icons, outcomes and 'sense of place' – are met later in the case study. Before reaching that point, this chapter takes a discursive route that is unwaveringly qualitative in its orientation. It may be best considered as being experimental, explorative and speculative, with no clear model to follow, and in this it takes its cue from the advocacies of Sorensen and Carman (2009) about approaches to and methods in heritage studies. It shares with Robinson (2015) and his study of a defunct theme park the adaptation of an autoethnographic approach. For this study, the input is derived from one curious, thoughtful and contemplative observer-cum-tourist – the author – and it draws on cumulative experience from a number of solo visits to the Stringybark Creek site. That on-site study and most of the commentary is set in a context that is established by the important backgrounding section, which is concerned principally with contextual scene-setting, as it considers the pathway of tourism and its interactions with selected issues from positive psychology and quality of life (QOL), with controversial themescapes, and with heritage-based tourism and memories. Following those backgrounding considerations is the case study with its setting in the story of the bushranger-cum-outlaw Edward (Ned) Kelly, and the specific happenings associated with the gunfight at Stringybark Creek. Underpinning all of the considerations is the testing of the belief that, no matter how controversial and dark the story might be, the pathway of telling it through the medium of tourism will contribute positively to personal levels of enrichment and enlightenment, both specifically and generally. This is the value-adding contribution of tourism.

Background

Before launching into specific consideration of the matters of the Stringybark Creek episode, it will be useful to delve into a nest of ideas that have given shape to the approach to that study. Embedded in that nest are ideas about the nexus of tourism with positive psychology and QOL, with controversial landscapes, and with the memorisation of outcomes of visits to significant places.

Tourism, Positive Psychology and Quality of Life

The most contextual of these three is embedded in the rich and ever-increasing body of literature that is situated at the confluence of studies in positive psychology, QOL, well-being and tourism. One of the sources that had an impact on how this study was to develop was Moscardo's (2009) commentary on the need for attention to be given to understanding how tourism affects tourists, as a complement and companion to the more usual studies on the impacts of tourism on destinations and on the residents of those destinations. For example, what is the benefit for a tourist in visiting a heritage site? This is rather more than just an enquiry about motivation, and even about whether the motivation (and the effort to visit) was justified. It is about how that visit might have brought a change to behaviour, to beliefs, to value sets, and so on. If, as Filep and Pearce (2014: xii) have claimed, 'Tourism is arguably one of the largest self-initiated commercial interventions to promote well-being and happiness on the global scale', it has to be asked what those visits have achieved. There is some guidance on this that has been of benefit for this study, and it can be found in commentaries by Filep (2012) and Veenhoven (2000), in which there are references to both inner and outer realms of QOL. This is especially important for studies of this kind because they give countenance to tourism experiences that are internalised (by the tourist), such as an appreciation of what lies behind any story being told or any situation being described in it, and appreciations that might 'guide' future actions of the tourist. A useful conceptualisation to help with this is Landsberg's (2004, 2015, 2018) notion of 'prosthetic memories' and the benefit of retrospectivity, of 'being there' but also 'not being there', to experience what happened at the site without having to endure the consequences of that at the time the happening took place. A benefit of the prosthetic status in this case is that after the event, and with the advantage of hindsight and of contextual factors, the tourist is in a position to enrich and enlighten their judgements about the event. Among the commentaries on these backgrounding issues that have a particular usefulness for studies of this kind are Chen and Yoon (2019), Filep and Laing (2018), Sirgy and Uysal, (2016), Tiberius and Hall (2010), Uysal *et al.* (2016) and Waterman (2008), which are highly relevant to the specific needs of this case study.

Controversial Themescapes

It is one of the complicating factors for studies of the kind in this chapter to find that the happenings under consideration are neither joyful nor pleasure-seeking, and this almost inevitably draws the study into realms describable as dark, difficult, dissonant, controversial and disputed, with some studies being described as traumascapes. Ashworth and Tunbridge (2012), among others, have captured this situation with observations that for such places to make positive 'psychic returns', they must embrace some distinctiveness, uniqueness and potential for appeal that outweighs any negativity, ambivalence or dissonance about what happened, how it happened and who was involved. Nelson's (2020) useful summary review and the study of World War II memorabilia by Thomas *et al.* (2016) have highlighted these issues. Even so, tourist experiences at contentious and controversial sites will not be consistent, and any degree of enrichment and enlightenment will be influenced by what was being sought (if anything) and even the tourist's willingness to be receptive to ideas that may lie contrary to those already held. Samuels's (2015) study of Sicily's fascist past and Foot's (2009) study of Italian memories of past events are illustrative of this.

Heritage-based tourism, memories and 'prosthetic memories'

This leads conveniently to Sthapit and Coudounaris's (2017: 72) observation that 'tourist experience is a complex construct . . . and is inherently personal'; this raises two areas for consideration, with one focusing on what makes for a memorable personal tourism experience (MTE) and the other being Landsberg's (2004, 2015, 2018) commentaries about prosthetic memories.

Research about MTEs often positions the remembered experiences and recollections along a spectrum from positive to negative (Sthapit & Coudounaris, 2017; Tung & Ritchie, 2011). Although this is useful, most studies avoid disaggregating their findings in order to establish what the tourists will do with their better-informed position, or how the tourism pathway has been used for personal improvement, or whether the encounter has effected changes to personal attitudes and behaviour or led to the adoption of a new set of values after appreciating what the story has brought up for them. The companion issue is that of prosthetic memories; Landsberg (2004: 222) has made the point that 'prosthetic memories emerge at the interface between a person and a historical narrative about the past, at an experiential site. . . . [A]t this moment of contact an experience occurs through which a person sutures him or herself into a larger historical narrative'.

Behind her observation lies the belief that more than a simple learning about the past is achieved, with the encounter and interaction becoming more personal, more meaningful and possibly deeply felt – although in reality it is no more than a 'memory' of something that was not actually experienced at the time it happened. The principal characteristics of prosthetic memory may be summarised as follows:

- 'being there' without actually 'being there';
- being historically specific and almost certainly politically inclined;
- relying on memory as affective and sensory rather than being merely cognitive;
- being neither natural nor necessarily authentic, but rather being mediated through transmission and presentation (for Landsberg's case it is the cinema; for this study it is heritage tourism);
- 'serious memories' that have outcomes (such as behavioural change);
- being commodified – that is, they are not 'the real thing' but rather can be substituted, modified and exchanged;
- *not* being commodities or commodified images as in the form of 'capsules of meaning that spectators swallow whole, but rather the grounds

upon which social meanings are negotiated, contested, and sometimes constructed' (Landsberg, 2018: 149);
- being helpful, empathetic, if not always comfortable.

For this study, the usefulness of the conceptualisation of the prosthetic memories lies in their facility for the construction of bridges with (or to) the past, and for 'seeing' through the eyes of those from past periods – that is what a visitor to the Stringybark Creek may seek to do, and it is that which is embedded in the autoethnographic exercise conducted by this author and reported on through the case study.

Methodology

Fashioned to meet a particular challenge, and with needed informational updates, this chapter draws on and will contribute to a long-standing investigation of doctoral research into the telling of stories about folk heroes from history through the medium of heritage-based tourism. There is now a considerable core from which baseline information can be drawn and that has been largely suitable for this exercise, meaning it has not required the discovery of extensive new baseline information. Rather, this exercise has worked from the existing core of baseline information currently available, that is, in any case, continually updated from scanning published sources (commercial texts, government agency documents, theses, academic papers, among other types), visits to libraries and museums where pertinent documents and artefacts are on deposit, and a commitment to a schedule of regular visits to sites and places associated with the stories being followed.

Every project working from the core information baseline has been approached methodologically from a hybridised investigative and interpretive process and framework that has been derived from the study areas of geography, history, literature and semiotics. This approach is consistent with the general strategy advocated by Moustakas (1990: 43), who has suggested that 'each research process unfolds in its own way . . . [and that] any course that a researcher's ingenuity is capable of suggesting is an appropriate method for scientific investigation'. It is also a nice fit with the general recommendations for support for cross-disciplinary research (Darbellay, 2016).

For this study two changes were made to what has become a standard process. The first of these was going beyond the core disciplines of geography, history, semiotics and literature studies and venturing into sources that were rooted largely in psychology. This was necessary so as to gain at least a toehold grasp on the linkages between tourist motivations and the possibilities of pathways of tourism to the outcomes of enrichment, enlightenment, well-being and 'authentic happiness'. The second variation was essentially for the sake of the case study. It involved refreshing fieldwork visits to three selected sites that held potential for this study; these were the siege site at Glenrowan, the Melbourne Gaol precinct and Stringybark Creek. All three sites were visited unaccompanied (by this author). The final decision was a preference for using the Stringybark Creek site, even though it was at one stage thought that all three could be used in this study; that idea was abandoned as being logistically impractical. An autoethnographic approach was taken to the fieldwork, involving journal entries, photography, participant observation, virtual participation and self-reflection, drawing on the experiences reported by Robinson (2015) and Wall (2016).

Case Study

One of the core issues for this study is that it is preferable to approach it with at least a reasonable grasp of the main elements of the story of Edward (Ned) Kelly, the Australian bushranger, outlaw and convicted murderer. Without some prior knowledge of the Kelly story, and especially of the exploits of the period 1878–1880 (which has come to be referred to as 'the Kelly Outbreak'), visits to the three important sites of Glenrowan, Melbourne and Stringybark Creek will leave only a surface-level impression of the significance of the man, of his story, of the events in which he was involved, and of the background circumstances. Such a visit would be little more than a day outing, perhaps climaxing with the acquisition of souvenirs and the collection of story-related memorabilia. The significance of the Kelly story – or at least the dramatic period of the Kelly Outbreak – to the history of Australia, the backgrounding elements (what led to and followed the events at Stringybark Creek), and what meanings it has for developing towards an understanding of the origins of Australianism and of the various sacrifices that were made by various individuals and social groups to advance the cause of 'Australia fair' deserve more than a casual acquaintance with the site at Stringybark Creek. This entire chapter is predicated on this belief, and on the proposition the medium of heritage-based tourism has an important role to play. As this section will portray, a visit to Stringybark Creek will not fit the specifications for a pleasure-seeking day out; hedonism is not a yardstick here.

The story of Edward (Ned) Kelly

The story of the Kelly family, their close friends, and those who became identified as their supporters and sympathisers is set in the second half of the 19th century as British colonial rule was being replaced progressively by state-based jurisdictions with their own codes of law and justice, and by the momentum towards the achievement of independent statehood by Australia on 1 January 1901. It was a period of changes in cultural, economic, political and social influence, a period of ongoing immigration (including from Ireland), a period of the occasional burst of gold mining activity, town growth and urban expansion (Jones, 2008; McQuilton, 1979). As so often happens in situations such as these, it was the rural communities that were among the last to benefit from the positive elements of the changes, and this was one of the important contributions to the civil disturbances (including, for example, the Eureka Stockade rebellion, 1854) in which the Kelly family members and their associates were often engaged, most particularly in confrontations with the police and the judiciary. It was not necessarily the case that the Kellys were especially confrontational, but court and police records contain inventories of arrests, charges, convictions and sentences that some observers consider to have been harsh and disproportionate. What all of this achieved, of course, was almost continuous notoriety, so that the story of Ned Kelly has become the stuff of legend, setting him high in the pantheon of Australian bushrangers and outlaws (Seal, 1996, 2011). It has been suggested that Ned Kelly 'offered more than bushranging. . . . [H]is courage and strength spoke for the frontier. . . . [I]n his daring and fatalism he chose both a short and merry life . . . resistance more than crime . . . protests at injustice and hints at a republic in north-east Victoria' (Gammage, 1998: 362). His activities have attracted presentations through a wealth of publications, dramas, musicals, films and artwork of various kinds, and the 'real' circumstances of the story have a long history of controversy, dispute and dissonance.

The Ned Kelly episodes

Kelly's life was short (1854–1880), and it came to a climax through the series of events commonly referred to as the Kelly Outbreak (1878–1880). The events of the 'outbreak' were precipitated by a rapidly deteriorating relationship between the Kelly family and the local police force, and the drama that extended across a period of two years and 16 days (October 1878–November 1880) began with the Stringybark Creek episode and included bank robberies at Euroa and Jerilderie, the murder of a police informant, and a siege at Glenrowan, climaxing with the execution of Ned Kelly in Melbourne Gaol (November 1880). There was a persistent undercurrent of a culture of mutual distrust, disrespect, and even loathing between the Kellys and the police force and the judiciary. Ian Jones (2003: xi) has described the situation as follows:

> Some saw the Kellys as arch-villains threatening the precarious stability of British rule and law in the self-governing colony of Victoria. To others the four members of the [Kelly] Gang were champions of the underdog, a focus for all the frustration and discontent of rural battlers scarred by bad seasons and a tottering economy.

Prima facie, the case is simple: Kelly robbed banks, murdered three policemen and a police informant, and was engaged in the final siege at

Glenrowan. But there is a backstory to that frontstory, and it has three components; the first of these, the most straightforward, is that Kelly and his associates engaged in unlawful acts and, therefore, should have expected the full force of the law to be applied to them. The second is that (as the previously used quotation from Gammage suggests), Kelly was supported because of his courage, for promoting the cause of a 'fair go' for frontier communities, for staring down the intimidatory tactics of the police, and for at least floating his ideals for a 'new' republic. The third leans on the perspective that the police that were involved in the confrontations were killed in action whilst doing their duty and obeying orders and so they were later lauded for their bravery. Each of these components of the backstory persistently influence interpretations of what happened at Stringybark Creek, and two of them emerge as notable themes for any meaningful contemplations across the site.

Stringybark Creek: The site and its tourism potential

A synopsis of the Stringybark Creek site situation is set out in Figure 10.1 and Table 10.1.

The site is, in a physical sense, spectacularly unspectacular. It is a typical regional bushland setting of various scrub grasses and eucalyptus trees, and sightlines in most directions are very short (except across the few clearings) because of the density and intensity of the landscape. There were few permanent structures there at the time of the confrontation between the four members of the Kelly gang and the four-man police patrol, and forest fires in the area and gradual deterioration have eroded since that time almost any vestige of structures and place markers. All presentations about the story rely on three sources: the record contained in the report made by the only policeman to have escaped from the scene of the gunfight; the records, sketches made, and photographs taken by the police party that was sent to the site to recover the bodies of the murdered policemen; and two of Kelly's own famous letters of grievance – the Cameron and Jerilderie letters. As there is not a lot of consistency across these sources, there have been long-running disputes about the precise locations of, for example, the police camp, the Kelly camp, what is sometimes referred to as the Kelly 'target' tree, the Kennedy tree where the third policeman died, the routes into and out of the camp site taken by Kelly and his associates and by the police, and even the geographical extent of the gunfight site. Rather than detract from the story, these inconsistencies and uncertainties imbue it with an aura of mystery.

So, what is there to draw tourists to the site? There is potential for probably three niche markets. One of these would be for tourists with a serious interest in various aspects of the Kelly story *per se*, with the visit functioning almost as a pilgrimage to Kelly sites. A second would be for tourists with an interest in sites that are linked to the series of events that

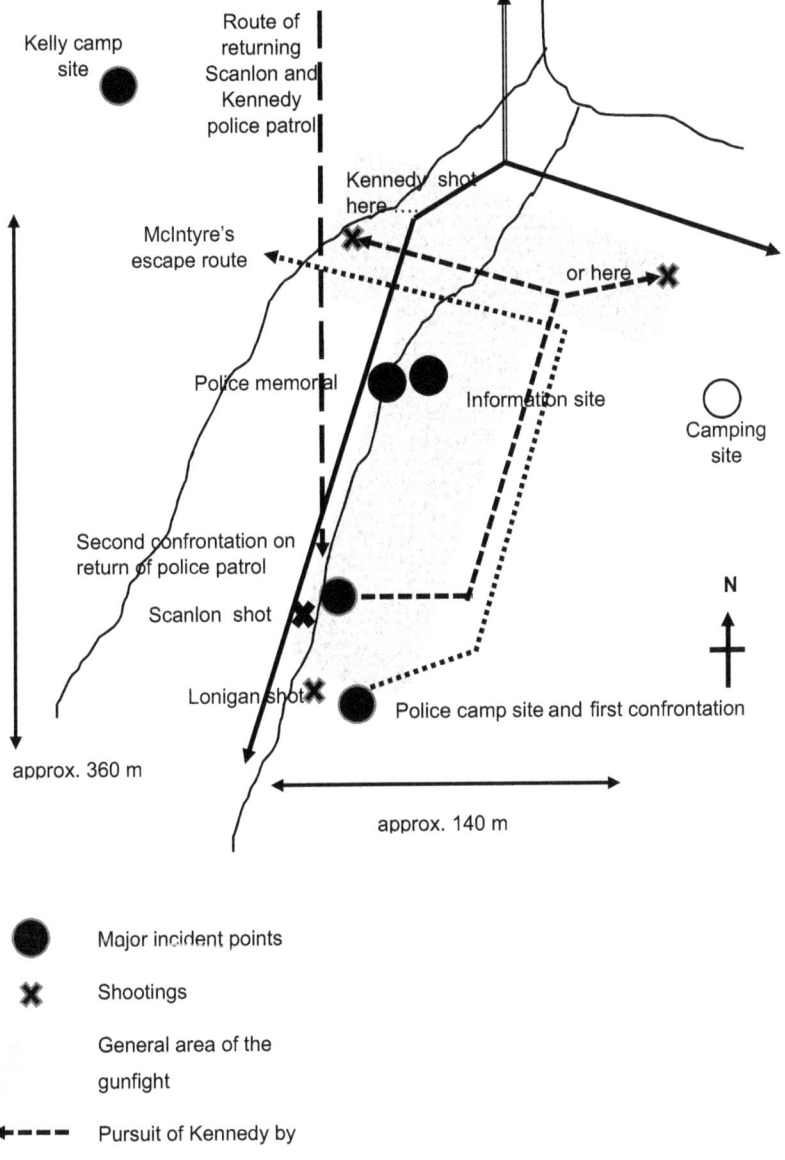

Figure 10.1 Stringybark Creek: approximate configuration of the action

were part of the 19th-century rebellion (including the Eureka Stockade episode in December 1854) that led on to the emergence of Australia as a sovereign nation. The third would be based around the theme of nostalgic and empathetic visits to acknowledge the bravery of the three policemen

Table 10.1 The significance of Stringybark Creek

Reason for its significance
The Stringybark Creek site marks the location of the police camp where the gunfight took place between the four members of the Kelly gang and a police patrol assigned the responsibility of searching for and apprehending Ned Kelly for an alleged attack on another policeman (Fitzpatrick) at the Kelly family homestead at Greta.

Location
Whilst the precise extent of the camp site and the gunfight remain uncertain, judgements have been made about where the action took place by merging police notes, a set of photographs taken at the time the bodies were being recovered, and commentary in Kelly's Cameron and Jerilderie letters. There is a consensus that the general location is about 10 km along a gravel track that deviates near Tomie from a minor country road that links the townships of Mansfield and Whitfield into the Toombullup Plantation and State Forest.

Basic circumstances
The gunfight site is set amidst bushland typical of the region: sassafras scrub, spear grass, stands of blackwood, mountain ash, and other eucalyptus trees. Small clearings and former camp sites form a patchwork across a network of small streams running in shallow ravines. Historically the area has been committed at various times to forest cultivation, logging and small-scale timber milling, clearance for cattle grazing, and some small-scale gold prospecting. Forest fires have been a recurring hazard.

The gunfight
This took place across a period of several hours and in two phases. In the first, one policeman (Lonigan) was shot at the camp site; in the second, two policemen (Scanlon and Kennedy) were shot as they returned to the camp from a patrol. Five days after the event the members of the Kelly gang were declared to be 'outlaws' (Felons' Apprehension Act, 1878)

Tourism potential (see Figure 10.1)
The precise locations of key elements – the police and Kelly camp sites, the direction of arrival of the police patrol, the site of Kennedy's death, the direction of the escape routes (of the surviving policeman, McIntyre, and of the Kellys after the siege – are in a state of constant controversy. And this is fuelled by the diversity of viewpoints, including the Kelly commentaries and those of family descendants, the police records, the descendants of the four murdered policemen, Kelly sympathizers, and a succession of private and commercially sponsored investigations. All of this adds to the mystery, to the legacy, and to the diversity of interpretations that can each hold their own meanings. The outcome of this is that the site, which nevertheless has significance for Australian history, can be considered as marking, for example,

- the beginning of the end of the activities of one of the legendary bushrangers;
- one of the key sites of the Kelly Outbreak and one of the culminating rebellious phases of the progression towards Australian independence and nationhood; and
- evidence of police bravery and devotion to duty.

Despite its significance for all or any of these, it is a primitive and remote site that is not supervised or serviced with water and electricity; commercial services are located more than 40 km away (in Mansfield); there is an adjacent but unsupervised and unserviced camp ground; the principal access points are from the camp site and from a gravel track that has to be shared with logging vehicles.

that were killed at the site in the gunfight, and this would be coupled with visits to the police monuments in Mansfield.

Any tourist-related potential is likely to be impeded by a number of access problems and servicing deficiencies. Among its locational problems are the unavoidable facts that the site is remote and that even within a radius of 40 km there are only a few small townships. The site is never closer than about 70 km from any highly trafficked major highway. It is approximately

250 km (say three hours' driving time) north-east of Melbourne and 700 km (say eight hours' driving time) south-west of Sydney, the two cities that have the largest pool of potential visitors. The site's remoteness is hardly conducive to spontaneous and serendipitous visits, and there are no commercial services within a 40 km radius to support visitation levels. The nearest township is Mansfield; unfortunately, it has a poor relationship with Stringybark Creek because Mansfield was the hometown and base of the three policemen who were killed in the gunfight.

It is therefore a site suited only to purposeful tourism.

Investigative Framework

For this study, an investigative framework was devised to help identify the important influences that may contribute to the shape and substance of the story to be told of the events at the Stringybark Creek site. In creating the framework, most attention was given to those influences that had the capacity to help conjure the elements of meaning that can be associated with the site and that can be associated with contemplation and reflection.

As there are so few certain markers of any of the events on the site, the best possibility was always going to be a concession that a particular position would be 'that which is most likely'; that is the strength of the comment made on the 'official' plaques around the site. High levels of accuracy and authenticity are unlikely; it is a site where imagination, speculation, and intelligent guesswork will be rife. The bushland setting conjures an atmosphere of mystery and intrigue, with uncertainty about where the police and Kelly camps might have been positioned, how the drama might have unfolded, where at any time the protagonists might have been situated, who moved in which direction and from which position to another, who did what and in what sequence, where the action was situated vis-à-vis the camp sites, the precise positions where the three policemen were shot, the direction in which the surviving policeman fled, the various firing sightlines of the participants in the drama, what the Kellys did and where they went after the gunfight, and so on and so on. As none of this action was recorded accurately at the time, as none of it was scripted, as there were no stage directions, and as most of the artworks, dramas, films, and non-fiction accounts work to their own convenient spatial agenda, for many of the happenings any one tourist's intelligent assessment could be as valid as that of another. Even expert studies of camp positions have failed to reach an accord sufficient for them to be recorded on 'official' heritage registers.

All this confusion leaves the potential tourist with latitude for a thought-provoking, challenging, but also enriching and enlightening encounter with the site. Drawing on a reasonable personal knowledge of the story and on a recall of previous investigations at the site, this author adopted a reflexive approach that was shaped by an already formed idea of what to look for and where to look. An outcome of the useful backgrounding and prior

Purpose: To identify possible tourism themes from circumstances at Stringybark Creek

Step 1: Assess baseline information position:

- Prior knowledge and already available information sources
- Revisions, updates, adjustments as necessary

Step 2: Assess likely influences on the scope and content of tourism themes:

ICONS	OUTCOMES	SENSE OF PLACE
Personal: Ned Kelly as an individual; Kelly gang as a collective; Kelly sympathisers as the support group **Place:** gunfight site	**At the site:** confrontation, gunfight, deaths **Trigger point:** Outlaws **Momentum:** Towards nationhood, identity	**Location:** remote **Nature of the place:** mysterious, surreal, disturbed by action **Embedded reactions and sentiments**

Step 3: Autoethnographic study to refine and differentiate evidence, as a step towards identifying supportable themes

Step 4: Using tourism as the pathway to add value to a story by crystallising possible themes, with each predicated on
- thoughtful images and memories;
- converting those images into a new understanding, a new attitude and behaviour, a new set of values and beliefs;
- converting those into positive and sustainable outcomes, which in this case would include the following:

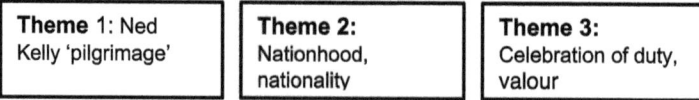

Theme 1: Ned Kelly 'pilgrimage'	Theme 2: Nationhood, nationality	Theme 3: Celebration of duty, valour

Figure 10.2 Using tourism as a pathway to enrichment and enlightenment

familiarity with the site was the construction of the framework that is Figure 10.2. It has four components through four steps:

(1) a baseline composed from prior knowledge and previous studies;
(2) a component of adjustments, revisions, and updates (to the previously existing baseline) involving a differentiation between the basic/

frontstory, the backstory, and the contextual circumstances; site-specific updates from site visits; and the input from a specially conducted autoethnographic study;
(3) a differentiation of such influential matters as iconicity (drawing from Jansen-Verbeke, 2016; Pearce *et al.*, 2003; and others) as outcomes, and 'sense of place' (drawing from Lappin, 2015; Park & Stephenson, 2007; Schmitz, 2016; and others);
(4) and a triptych of three tourism themes, in which any one has its own tourism potential, and any combination of two or all three could sustain a focused tourism strategy.

A form of autoethnographic study was undertaken to form an interpretational bridge to the identification of possible focused tourism themes. The focus of this was semiotically strong, as what was being looked for were markers across the site that would meet the benchmark challenge of Urry (1992) about distinctiveness and a reasonable likelihood that what could be identified would have touristic attraction. From the study three themes were readily identifiable:

(1) the pilgrimage theme: a theme that incorporated the Stringybark Creek site in a pilgrimage that with other sites would contribute to telling the Ned Kelly story; this would recognise his iconic and legendary status;
(2) the nationhood theme: a theme that linked the Stringybark Creek site with the siege site at Glenrowan as constituents of the 19th-century movement that led eventually to the creation of the 'new' nation of Australia and the forging of its distinctive identity;
(3) the valour theme: a theme that viewed the Stringybark Creek episode as one at which the police distinguished themselves with commitment, and with bravery in confronting the members of the Kelly gang, so that the site becomes a memorial to the action of the police rather than the Kellys; this theme would link this site to the nearby township of Mansfield in which there are monuments to the police.

Although the focus here is on different motives and subject matter, this repertoire of three themes adopts a similar opportunistic approach to that taken by Edensor (1998) in his speculations about the Taj Mahal.

Throughout the study the underpinning intention has been to follow Landsberg's (2004) conceptualisation of 'being there without being there' and to approach the site with a serious intent to use the touristic experience as a means to engage with history.

Conclusion

From the very beginning of this chapter, it was made clear that what was about to be considered would not be an easy fit with conventional expectations of tourism as a pathway to hope, happiness, and other laudable aspirations of 'the good life'. In fact, a principal and nagging concern throughout this chapter has been how a situation that was clearly situated on the dark side of the tourism spectrum could become a vehicle for transporting the tourism experience to anything positive. This was the challenge, and the aim was to confront it by speculating that from serious and thoughtful contemplation of particular happenings at particular sites could come attitudes and values that might bring about a repositioning of even the darkest of events so that, through the medium of tourism, there could be positive and sustainable outcomes. One noteworthy outcome that was achieved through this process in this study was the revelation that the same set of dark circumstances could be exposed through tourism in differently themed expressions; interpreting the rich domain of 'evidence', three different themes were identified, one with its focus on a form of pilgrimage to Ned Kelly sites, another on the emergence of Australian nationhood, and a third on the celebration of the duty and valour of the Victoria police. In order to progress from the basic thought of the prospect of more than one dominant theme to the creation of a repertoire of at least three themes, the information base used for this author's previous Ned Kelly–related studies, and from prior reading of primary and secondary sources about the Kelly story, was updated and re-formed through a simple autoethnographic process that included further visits to the Stringybark Creek site and inputs from a rather untidy process of note-making, sketching, photography and reflective thinking while there.

The study has been driven by two claims that were made early in this chapter. These were:

(1) that the study would be 'heavily predicated on experiencing images and drawing on memories and then converting those images into a new understanding, a new attitude and behaviour, and a new set of values of and for the story being told'; and
(2) that, 'no matter how controversial and dark the story might be, the pathway of telling it through the medium of tourism will contribute positively to personal levels of enrichment and enlightenment'.

Enrichment and enlightenment were expected to be the positive outcomes; in summary, the aim was to discover a way of value-adding to a story from history told through the medium of tourism, so that even if it didn't exactly encapsulate 'hope and happiness' it would have the potential to contribute to them. One of the early steps was to negotiate the intermixture of the intricacies of tourism, positive psychology, QOL, controversial themescapes and a range of dimensions about memories. Then, 'being there without being there' was set as the foundation stone of the study because it was expected (or hoped) that what could be learned from the circumstances encountered at the site would help towards an understanding of what really happened – not only the what but also some unravelling of the how and importantly the who, the where, the when and the why.

So, how might the presentation through tourism of a dramatic happening in the Ned Kelly story contribute to any measure of hope and happiness? Sirgy and Uysal's (2016) commentary offered a useful starting point. They presented three dimensions of measurement. One of these differentiated between psychological happiness, prudential happiness and perfectionist happiness; the second differentiated according to whether the affect is positive or negative; and the third focused on whether or not there was a reward (or benefit). A detailed assessment of these differentiations in the specific case used in this chapter is beyond the scope of this study, even though, as an exercise, it would be instructive. So, for the study in this chapter the encounter with the 'dark-ish' subject of the Ned Kelly story as a whole, and the Stringybark Creek episode in particular, is most likely situated comfortably within what Seligman (2002) has described as an 'engaged life'. This is one that challenges personally held views, abilities, and skills (and leads to improvement), that 'serves a higher purpose' and is of benefit to other people, and that is ethically and morally defensible. For the Kelly-based case study, and for others like it, this most likely means that superficiality of investigation, knowledge-building and interpretation would be responses best suited to pleasure-seeking, sight-seeing and serendipitous tourism, while facing up to the contentious challenges embedded in the Kelly story and at the Stringybark Creek site both need and deserve a serious encounter that befits the status of 'being there without being there' – searching for meanings (the why), learning lessons (the what if), generating positive emotions (such as awe, compassion, gratitude, honour and respect), establishing cultural and historical reference points (such as landmarks and story markers), and more.

A multiplicity of possible outcomes in terms of tourism themes is a bonus that accrues from giving a site a more intensive investigation and interpretation than would be the usual case with pleasure-seeking, hedonistic tourism. For heritage-based tourism this value-adding benefit has possible flow-on consequences for the cultural, economic, political and social well-being of a tourism-targeted site and the area in which it is situated. For the study in this chapter, the focus has been on a contentious story, set amid shades of darkness. But the principle (if that is what it is) of multiple opportunities that can flow from giving any site deep and meaningful thought can be used away from the shades of darkness. For example, Edensor (1998) has demonstrated in his study of the Taj Mahal that there are three stories (represented by the symbolic structure) there: one that refers to the colonial past, another that references Islamic faith and power, and a third that is a monument to love. The lesson? It is the depth and breadth of the thinking about the subject site that will reflect the capacity and direction of tourism as a pathway.

References

Ashworth, G. and Tunbridge, J. (2012) Heritage, tourism and quality-of-life. In M. Uysal, R. Perdue and M. Sirgy (eds) *Handbook of Tourism and Quality of Life Research* (pp. 359–371). Dordrecht: Springer.

Chen, C. and Yoon, S. (2019) Tourism as a pathway to the good life: Comparing the top-down and bottom-up effects. *Journal of Travel Research* 58 (5), 866–876.

Darbellay, F. (2016) From disciplinarity to postdisciplinarity: Tourism studies dedisciplined. *Tourism Analysis* 21, 363–372.

Edensor, T. (1998) *Tourists at the Taj: Performance and Meaning at a Symbolic Site*. London: Routledge.

Filep, S. (2012) Positive psychology and tourism. In M. Uysal, R. Perdue and M. Sirgy (eds) *Handbook of Tourism and Quality of Life Research* (pp. 31–50). Dordrecht: Springer.

Filep, S. and Laing, J. (2019) Trends and directions in tourism and positive psychology. *Journal of Travel Research* 58 (3), 343–354.

Filep, S. and Pearce, P. (eds) (2014) *Tourist Experience and Fulfilment*. Abingdon: Routledge.

Foot, J. (2009) Divided memory: Theory, methodology, practice. In *Italy's Divided Memory* (pp. 1–29). New York: Palgrave Macmillan.

Gammage, B. (1998) Kelly, Edward 'Ned'. In G. Davison, J. Hirst and S. McIntyre (eds) *The Oxford Companion to Australian History* (pp. 362–363). Melbourne: Oxford University Press.

Hartmann, R. (2014) Dark tourism, thanatourism, and dissonance in heritage tourism management: New directions in contemporary tourism research. *Journal of Heritage Tourism* 9 (2), 162–182.

Jansen-Verbeke, M. (2016) Tourismification of cultural landscapes: Synergies between tangible and intangible heritage resources. In *Proceedings of TCL2016 Conference* (pp. 276–281). INFOTA, Budapest, 12–16 June. Conference proceedings.

Jones, I. (2003) Introduction. In J. Corfield (ed.) *The Ned Kelly Encyclopaedia* pp. i–xv Melbourne: Lothian Books.

Jones, I. (2008) *Ned Kelly: A Short Life*. Melbourne: Lothian Books.

Landsberg, A. (2004) *Prosthetic Memory*. New York: Columbia University Press.

Landsberg, A. (2015) *Engaging the Past*. New York: Columbia University Press.

Landsberg, A. (2018) Prosthetic memory: The ethics and politics of memory in an age of mass culture. In P. Grainge (ed.) *Memory and Popular Film* (pp. 144–161). Manchester: Manchester University Press.

Lappin, L. (2015) *The Soul of Place*. Palo Alto: Traveler's Tales.

McQuilton, J. (1979) *The Kelly Outbreak: The Geographical Dimensions of Social Banditry*. Carlton: Melbourne University Press.

Moscardo, G. (2009) Tourism and quality of life: Towards a more critical approach. *Tourism and Hospitality Research* 9 (2), 159–170.

Moustakas, C. (1990) *Heuristic Research: Design, Methodology, and Applications*. Newbury Park: Sage.

Nawijn, J. and Filep, S. (2016) Two directions for future tourist well-being research. *Annals of Tourism Research* 61.C, 221–223.

Nelson, V. (2020) Liminality and difficult heritage in tourism. *Tourism Geographies* 22 (2), 298–318.

Park, H. and Stephenson, M. (2007) A critical analysis of the symbolic significance of heritage tourism. *Intenational Journal of Excellence in Yourism, Hospitlaity and Catering* 1 (2) 39-66 See http://respository.usp.ac.fj/8692/1/Article_49_HBMSU.pdf (accessed January 2021).

Pearce, P.L., Morrison, A.M. and Moscardo, G.M (2003) Individuals as tourist icons: A developmental and marketing analysis. *Journal of Hospitality and Leisure Marketing* 10 (1–2), 63–85.

Robinson, P. (2015) I remember it well: Epiphanies, nostalgia, and urban exploration as mediators of tourists' memory. *Tourism, Culture and Communication* 15, 87–101.

Samuels, J. (2015) Difficult heritage. In K. Samuels and T. Rico (eds) *Heritage Keywords* (pp. 111–128). Boulder: University Press of Colorado.

Schmitz, H. (2016) Atmospheric spaces. *Ambiences* (open access edition). See http://journals/openedition.org/ambiences/711.

Seal, G. (1996) *The Outlaw Legend*. Cambridge: Cambridge University Press.

Seal, G. (2011) *Outlaws in Myth and History*. London: Anthem Press.

Seligman, M. (2002) *Authentic Happiness: The New Positive Psychology to Realize Your Potential for Lasting Fulfilment*. New York: Free Press.

Sirgy, M. and Uysal, M. (2016) Developing a eudaimonia research agenda in travel and tourism. In J. Vitterso (ed.) *Handbook of Eudaimonic Well-Being* (pp. 485–495). Cham: Springer.

Sorensen, M. and Carman, J. (eds) (2009) *Heritage Studies: Methods and Approaches*. Abingdon: Routledge.

Sthapit, E. and Coudounaris, D. (2017) Memorable tourism experience: Antecedents and outcomes. *Scandinavian Journal of Hospitality and Tourism* 18 (1), 72–94.

Stone, P. (2006) A dark tourism spectrum: Towards a typology of death and macabre related tourist sites, attractions and exhibitions. *Tourism* 54 (2), 145–160.

Thomas, S., Seitsonen, O. and Koskinen-Kolvisto, E. (2016) Dark tourism. In C. Smith (ed.) *Encyclopedia of Global Archeology*. Berlin: Springer AG., https://doi.org/10.1007/978-3-319-51726-1_3197-1.

Tiberius, V. and Hall, A. (2010) Normative theory and psychological research: Hedonism, eudaimonism, and why it matters. *The Journal of Positive Psychology* 5 (3), 212–225.

Tung, V. and Ritchie, J. (2011) Exploring the essence of memorable tourism experiences. *Annals of Tourism Research* 38 (4), 1367–1386.

Urry, J. (1992) The tourist gaze revisited. *American Behavioural Scientist* 36 (2), 172–186.

Uysal, M., Sirgy, M., Woo, E. and Kim, H. (2016) Quality of life (QOL) and well-being research in tourism. *Tourism Management* 53, 244–261.

Veenhoven, R. (2000) The four qualities of life. *Journal of Happiness Studies* 1 (1), 1–39.

Wall, S. (2016) Toward a moderate autoethnography. *International Journal of Qualitative Methods* (January–December), 1–9.

Waterman, A. (2008) Reconsidering happiness: A eudaimonist's perspective. *The Journal of Positive Psychology* 3 (4), 234–252.

11 Navigating the New Normal: Restorative Tourism Experiences during Times of Crisis

Sera Vada and Noel Scott

Design of restorative experiences for domestic tourists is an emerging topic of research and practice. This chapter draws upon a tourist well-being framework (Vada et al., 2020) in order to provide recommendations for developing restorative tourism experiences during the COVID-19 pandemic. Studies have shown that tourism and travel is beneficial to enhancing happiness and well-being. However, the COVID-19 pandemic has disrupted travel behaviour with the closure of travel borders and stay-at-home restrictions. The pandemic also has social and lifestyle implications, greatly changing the way we live through isolation and, notably, physical and social distancing from family and friends. With international borders closed for the foreseeable future, tourists are travelling within their own country for restorative purposes. The findings of this chapter highlight the significance of interaction with the natural environment as it helps to address isolation and enhances physical health and general good health. It also emphasises the importance of social relationships in tourism consumption which equally enhance well-being and quality of life. This chapter therefore assists tourism operators in developing the capacity to create, shape, and stimulate restorative experiences that will enhance the psychological and social well-being of domestic travellers.

Introduction

Tourism is a driver of the world's economy, contributing 10.3% of global GDP and 330 million jobs around the world (WTTC, 2019). However, travel also plays an important role in people's lifestyles, social harmony and maintaining psychological health as it locates people in novel situations and provide benefits such as broadening travellers' view of the world as well as altering their attitudes and allowing them to face challenges (Kottler, 1997). The impact of the COVID-19 pandemic has not only led to a standstill for tourism and travel but has also greatly changed the way we live through isolation in the form of physical and social distancing from family and friends. Notably, during the pandemic, many people have experienced a heightened state of anxiety and have struggled to manage the uncertainty surrounding the possible spread and impact of COVID-19. In Australia, whilst international borders remain closed, tourism slowly increased from day trips to intrastate and interstate travel (although Western Australia remained closed to interstate travel for two years). In 2019, Australians took 6.4 million leisure trips overseas; however, the COVID-19 pandemic confined trips to domestic travel until early 2022 when international borders reopened.

Tourism and travel are beneficial as they can enhance happiness and well-being. Studies have shown that people often feel happier, healthier, and more relaxed after a holiday (Nawijn *et al.*, 2010). However, Nawijn (2011) found that happiness after holidays is mostly short-lived because tourism experiences are not considered special anymore as travel has become a much more integral part of life and tourism has become more accessible in everyday life. This chapter interprets well-being from the viewpoint of positive psychology, which operationalises well-being through two approaches: hedonia and eudaimonia. The concept of hedonia focuses on positive emotions, happiness and pleasure, while eudaimonia focuses on personal growth and optimal functioning (Huta, 2013). An important distinction between the two approaches is that hedonia is about feeling good whilst engaging in an activity, whereas eudaimonia can result from activities that are not particularly pleasant at the time but may have delayed positive effects that occur well after a trip. These may include increased skill level or reaching a goal (Huta, 2013).

Tourism offers products and services to consumers where one can experience either hedonic or eudaimonic well-being. Hedonic views of well-being are common in the tourism literature, with happiness and pleasure being seen as the ultimate goal. However, eudaimonic well-being, whereby tourist experiences provide meaning that involves deep satisfaction as well as learning, personal growth and skill development, is a growing area of interest for tourism scholars (Pearce & Packer, 2013). It has been argued that hedonic tourism products and services are usually categorised by excessive behaviour such as eating and drinking, whereas tourism products and services such as walking trails or cycling can help tourists to realise benefits in their own health and improve eudaimonic well-being (Pyke *et al.*, 2016). The academic tourism literature has examined a number of tourist experiences that may influence well-being, including wellness and spa tourism (Voigt *et al.*, 2011), religious and spiritual travel (Chamberlain & Zika, 1992), sport tourism (Filo & Coghlan,

2016) and volunteer tourism (Crossley, 2012). There is a call in the literature for tourism and hospitality scholars to continue to develop research agendas on tourist well-being in travel and tourism as the focus on the quest for self-development and transformation has become a central concern of Western society (Sirgy & Uysal, 2016).

Existing studies suggest that natural, social, and built environments provide the attributes necessary for a restorative experience and thus have the ability to create a sense of peace and calm that enables people to recover their cognitive and emotional effectiveness (Hartig *et al.*, 1991; Kaplan, 1995; Korpela & Ylén, 2007). Studies have highlighted the natural environment's restorative powers in research on rural tourism (Sharpley & Jepson, 2011) and wildlife tourism (Curtin, 2009). Features of the built environment that have been studied as restorative environments within the tourism context include museums (Packer, 2013) and urban parks (Chiesura, 2004). There are also studies of simulated natural environments such as zoos and aquariums (Falk *et al.*, 2007). What is significant now, more than ever before, is the need for research that focuses on defining and designing tourist experiences as restorative experiences post-COVID-19 pandemic and as a way to manage in times of crisis.

There are a number of significant reasons supporting further research on travel and well-being. On a global scale, this research is aligned with United Nations Sustainable Development Goal 3, which is to promote good health and well-being. Currently, the world faces a global health crisis unlike any other with COVID-19 spreading human suffering, destabilising the global economy and having devastating effects on the lives of billions of people around the world. The field of positive psychology is an area of study that seeks to highlight the role of positive emotions, character strengths and positive institutions that serve human happiness and well-being. The tourism industry can serve as a positive institution during the pandemic by navigating the new normal and providing or enhancing restorative tourist experiences for travellers. This study provides a model of tourist well-being developed by Vada *et al.* (2020), which outlines three main areas of tourist well-being: (1) antecedents or triggers of tourist well-being; (2) episodes or consumption contexts of tourist well-being; and (3) consequences or benefits of tourist well-being. As this model has yet to be tested with empirical data, this model is used to highlight lessons on developing restorative tourism experiences during the COVID-19 pandemic. In the developing literature on well-being, the extent to which specific experiences within tourism or travel contexts contribute to tourists' well-being remains unclear. The chapter will then present different ways that local or regional holidays may contribute to restoration and well-being through natural, social and built environments.

Literature Review

Tourist well-being framework

Vada *et al.* (2020) found that tourist well-being has been predominantly examined as an outcome variable from tourist experiences. There is an evident need to link tourist well-being to practical outcomes that would be beneficial to tourism managers and marketers, specifically in the areas of behavioural intentions and destination attachment. The authors developed a tourist well-being framework that focused on three main areas: (1) antecedents or triggers of tourist well-being; (2) episodes or consumption contexts of well-being; and (3) consequences or benefits of tourist well-being (see Figure 11.1).

Antecedents refer to triggers that may influence tourist well-being, such as positive emotions, character strengths, mindfulness, engagement, relationships, meaning and accomplishment. There are a few positive psychological theories that may explain how these variables influence tourist well-being. One example is the broaden-and-build theory of positive emotions (Fredrickson, 2004), which claims that certain positive emotions broaden people's momentary thought–action repertoires and build their enduring physical, social and psychological resources. Previous empirical research has found that positive emotions can help people to cope with adversity (Miao *et al.*, 2013; Mitas *et al.*, 2012; Tugade & Fredrickson, 2004) and also contribute to the improvement of cardiovascular health following negative emotions (Fredrickson *et al.*, 2003). The PERMA model of well-being (Seligman, 2004) has also been claimed to contribute to longer-term well-being through its five building blocks of positive emotions, engagement, relationships, meaning and achievement.

Antecedents *Triggers of tourist well-being*	Episodes *Consumption contexts of tourist well-being*	Consequences *Benefits for tourism marketing and management*

Variables
• Happiness
• Positve emotions
• Savouring
• Character strengths
• Gratitude
• Humour
• Mindfulness
• Engagement
• Relationships
• Meaning
• Accomplishment
 	
Positive psychology theories
• Broaden and build theory
• Flow theroy
• PERMA model
• Mindfulness theory
	• Natural environment and built environment
	  • Presence of silence
	• Social environment
	  • Acts of kindness
	• Tourist experiences
	  • Wellness tourism
	  • Yoga tourism
	  • Volunteer tourism
	• Countries
	  • Western countries
	
	Promotes tourist health and well-bring
	• Positive aging through social support
	• Belongingness to social world
	• Finding one's inner self
	• Ability to cope with the stresses of life
		• Revisit intentions
		• Positve word-of-mouth
		• Residents participation in value cocreation with tourists
		• Positive attitudes towards poverty alleviation and development issues

Figure 11.1 Conceptual tourist well-being framework (After Vada *et al.*, 2020)

Episodes refer to interaction with the natural and social environment that is present in tourism experiences and found to be linked to well-being – for example, museums (Packer, 2013), urban parks (Chiesura, 2004), zoos and aquariums (Falk *et al.*, 2007). It has also been found that experiences in nature can influence emotional well-being (Beckmann *et al.*, 1998) while acts of kindess influence emotional connections to peole within host communities (Filep *et al.*, 2017).

Consequences refer to the benefits of tourist well-being such as enhancing one's inner self and reinforcing the ability to cope with the stresses of everyday life. The benefits of experiencing constructs of well-being within tourism, such as the presence of silence (Dillette *et al.*, 2018), acts of kindness (Filep *et al.*, 2017) and meaning from holiday experiences such as wellness tourism and yoga tourism (Voigt *et al.*, 2011) have also resulted in behavioural consequences. These behavioural consequences have significant implications for tourism marketing and management as it relates to revisit intentions, positive word of mouth (WOM) and destination attachment. Existing studies from this review suggest that there is a relationship between tourist well-being and behavioural intentions. For example, Lin (2012) found that cuisine experience influenced psychological well-being, which consequently affected tourists' revisit intentions. The motivation and subjective well-being also affected the intention to revisit of hiking tourists (Kim *et al.*, 2015). Furthermore, Reitsamer and Brunner-Sperdin (2015) found that tourists' well-being had a positive impact on their intention to return and the desire to engage in positive WOM.

Tourism and restoration

Early studies on the tourist experience adopted a marketing and management disciplinary perspective whereby the construction of the tourist experience was based on the interaction between the individual tourist and components of the tourism system (Cohen, 1972; Leiper, 1979). In addressing the prejudice that tourism is often seen as an indulgent aspect of life and lacking in intellectual interest, Pearce (1987) argued that tourist behaviour and experiences was a stimulating and worthwhile topic for psychologists. As such, tourist experiences have been viewed from the perspective of social psychology when examining attitude changes (Ajzen & Fishbein, 1980; Um & Crompton, 1990), environmental psychology on the topic of restorative tourist experiences (Hartig *et al.*, 1991; Kaplan, 1995) and cognitive psychology on the topic of emotions (Brunner-Sperdin *et al.*, 2012; Ma *et al.*, 2013).

The concept of restoration can be traced back to early psychological theories of attention and clinical neurological studies of mental functioning, with considerable attention to the concept of restoration in the environmental psychology literature (Kaplan, 1995). Restoration focuses on the human need for rest and recuperation, recovery from mental fatigue and renewal of diminished capabilities (Hartig et al., 1991). Restorative experiences refer to the renewal of resources (physical, psychological and social) that have been depleted in meeting the demands of everyday life (Hartig, 2011). The attention restoration theory (Kaplan, 1995) claim that in order to fully recover from directed attention fatigue, it is important that the individual's attention is engaged involuntarily or effortlessly, rather than intentionally. The four components that have been identified as integral to a restorative experience are fascination (being engaged without effort); a sense of being away (physically or mentally removed from one's everyday environment); perception of extent (environment provides enough to see, experience, and think about to sufficiently engage the mind); and compatibility (providing a good fit with one's purposes or inclinations). Cimprich (1993) suggests that even short periods of time spent in a restorative environment can have significant effects on both cognitive capacity and quality of life. Ottosson and Grahn (2005) used the attention restoration theory on measures of restoration in residents in geriatric care comparing leisure time spent in a garden with leisure time spent indoors. In the tourism literature, the attention restoration theory is also widely used to explain preference for natural environments as tourists perceive nature as being associated with relaxation or escape (Pearce, 2012). Furthermore, the stress recovery theory has been used to explain how being in natural environments elicits stress-reducing responses (Ulrich et al., 1991).

Earlier studies in tourism literature suggest that the need or desire for restoration is an important motivation underlying engagement in tourism and leisure experiences (Crompton, 1979; Pearce & Lee, 2005; Uysal et al., 2012). A destination serves as a restorative environment that may improve emotional and cognitive functioning as well as imparting health and well-being benefits. The tourist experience is suggested to contribute to a restorative experience for tourists as they are physically or mentally removed from their everyday environment, which may lead to recovery from fatigue (Kaplan, 1995). Consequently, recent attention has been given to the therapeutic effect of holidays on mentally stressed individuals. Existing studies have tried to assess the link between taking a holiday and recovery, personal well-being and work-performance-related outcomes (de Bloom et al., 2010; McCabe et al., 2010; Nawijn et al., 2013; Pearce, 2012; Uysal et al., 2016). However, until recently they failed to identify the components of a holiday that restore and re-energise travellers. As a result, Lehto et al. (2017) developed an instrument for assessing the perceived restorative qualities of holiday destinations, which includes a 30-item, six-factor structure of destination restoration. This measure has since been verified by studies on topics such as on understanding the recovery process of Chinese tourists (Chen et al., 2017) and smartphones and holiday recovery (Kirillova & Wang, 2016).

As noted by Vada *et al.* (2020), episodes refer to interaction with the natural and social environment that are present in tourism experiences and found to be linked to well-being. The next section elaborates on restoration aspects in the natural and social environment that contribute to well-being.

Natural environment and restoration

Natural settings have been found to have a greater impact on attentional restoration than urban surroundings (Hartig *et al.*, 2003). Every year, over 8 billion people visit nature reserves such as beaches, islands, mountains and wilderness areas, generating an estimated revenue of $600 billion worldwide (UNWTO, 2019). Importantly, green spaces and blue spaces (e.g. gardens, parks, forests and oceans) are the most common natural destinations that influence health and well-being (Gascon *et al.*, 2015). It has been found that viewing a natural environment reduces stress faster than an urban environment (Ulrich *et al.*, 1991). Recent studies indicate that interaction with the natural environment may help address isolation and other negative mental effects resulting from the COVID-19 outbreak (Fiorillo & Gorwood, 2020). The study of perceived restorative environments is increasingly turning attention to soundscapes, smellscapes, tastescapes and hapticscapes, in addition to visualscapes (Dann & Jacobsen, 2003). Bunkše (2012) integrates all of these into the 'sensescape', defining it as a multisensorial space that is distinguished, understood, and valued by humans. A sensescape is where the environment and body meet (Rosen, 2018). The sensescape describes a multisensorial connection between the self, others, and the surrounding world, as a method of inhabiting and experiencing oneself and one's surroundings at the same moment (Jensen *et al.*, 2015).

It is reported that engaging with nature has direct benefits for physical health due to increased physical activity, which enhances fitness and general good health (Heung & Kucukusta, 2013). There are also significant studies that suggest that visiting natural areas is beneficial mentally through high levels of positive emotions and a sense of satisfaction with one's quality of life (Chen *et al.*, 2017; Ohe *et al.*, 2017).

Social environment and restoration

The concept of companionship in the social psychology literature involves protection from the emptiness and despair that is associated with loneliness (Buunk & Verhoeven, 2010; Rook, 1987). The benefits of companionship include psychological well-being, health and life satisfaction (Coleman & Iso-Ahola, 1993). Well-being and enhanced quality of life constitute central themes in positive psychology (Deci & Ryan, 2008; Smith & Diekmann, 2017). This research has transformed well-being into a 'buzz word' over the past decade, omnipresent in discourse about human daily life and activities.

The formation of interpersonal bonds is a fundamental and innate human need and is essential for well-being (Baumeister & Leary, 1995). Tourism scholars have highlighted companionship as a fundamental factor in determining leisure preferences and participation (Raymore, 2002). Previous research has also addressed the importance of social relationships in tourism consumption and, more importantly, that companionship can enhance travel experiences (Kim et al., 2010). It has been shown that travel experiences are significantly influenced by activities that affect emotions, such as talking to others (Olsson et al., 2013). Zhu and Fan (2018) identified a strong relationship between travel companionship, happiness, and meaningfulness. Those travelling with their spouse or partner, children, and other family members and friends are much happier than those travelling alone. Companionship also influences meaningfulness – higher levels are associated with all types of companions, other than colleagues. Matteucci et al. (2019) identified an association between shared travel experiences and positive outcomes such as strengthened friendship ties and personal growth.

Rather than being experienced and interpreted individually, tourism experiences are often group-based (Crompton, 1981). Previous researchers have linked travel companionship and memorable tourism experiences, tourist well-being and behavioural intentions. For example, Gregory and Fu (2018) concluded that people who travelled with family members enjoyed enhanced holiday satisfaction and overall well-being. A link was also identified between family size and happiness, with smaller families experiencing greater happiness and overall holiday satisfaction. Choo and Petrick (2015) found that whether travellers had memorable tourism experiences and would revisit the holiday location was positively influenced by travel companionship. Changuklee and Allen (1999) also found that behavioural intentions were highly related to traditional travel-related variables, such as destination attractiveness and travelling to the destination as a family tradition. Furthermore, Rosas de Gracia and Urbistondo (2020), found that family satisfaction was greatly influenced by travelling abroad and sharing activities.

Practical Implications

This review suggests some practical strategies for tourism marketers and managers in dealing with the effects of COVID-19 on people's health. Firstly, as tourist health and well-being is influenced by blue and green spaces, it is vital for tourism destinations to specifically promote and market their natural surroundings through visuals (images and text), as this can influence tourists' choice of a destination and revisit intentions. These promotional strategies can be aimed at people living in cities, as it has been reported that people in urbanised societies commonly believe that contact with nature provides them with restoration from stress and fatigue and significantly improves their health and well-being (Van Den Berg *et al.*, 2007).

Secondly, this review indicates that tourist well-being is more than just a physical activity and is also influenced by activities or experiences that involve enhancing social relationships, learning a new culture or developing a new skill. Wellness products and services within natural surroundings are generally perceived as luxurious and expensive; therefore, offering reasonable and cost-effective activities such as volunteering activities or homestay options in host communities would allow tourists to learn about a new culture and connect with the local people while relaxing in a natural environment. As a result, these initiatives would not only support the health and well-being of tourists and residents in host communities but would also have positive effects on environmental sustainability by promoting pro-environmental behaviour among visitors.

Finally, tourism marketers should expand general tourist or customer satisfaction questionnaires beyond satisfaction ratings. Other measures such as positive emotions, achievements and personal growth should be considered, as these variables are found to influence well-being (Vada *et al.*, 2019). This information would be significant because it can assist in the development of tourism products and services, which would maximise tourist satisfaction in ways that contribute to life satisfaction and tourists' quality of life. In addition, tourism marketers should promote staycations (during weekends) as a healthy and affordable option that enhances well-being during a shorter period of time.

Conclusion

Despite an increasing number of studies applying positive psychology concepts to the study of tourist well-being, the platform of tourism scholarship that directly explores tourist well-being is not well established (Filep & Laing, 2018). Pearce (2009) argues that positive psychology has the power to change the way that tourist behaviour is studied. This study has laid the groundwork by mapping what is known and yet to be known in the relationship between positive psychology and tourist well-being studies. By providing future directions for research, this chapter will enable tourism scholars to contribute further to this growing area and the emerging field of 'positive tourism' (Filep *et al.*, 2016). Through a conceptual framework, this chapter has also contributed to a better understanding of the antecedents, episodes and consequences of tourist well-being from the lens of positive psychology in tourism research. More significantly, this study has provided practical strategies by which tourist well-being can be utilised to generate optimal outcomes for tourism marketers and managers.

Thirdly, the restorative well-being benefits of a holiday are significant in tourism marketing as they can influence tourists' decision to visit a particular destination and, subsequently, behavioural intentions such as positive WOM and revisit intentions (Pyke et al., 2016; Reitsamer & Brunner-Sperdin, 2015; Sirgy & Lee, 2008). However, there are minimal studies examining whether well-being (both hedonic and eudaimonic) influences behavioural intentions and destination attachment. The practical implications of this research will assist tourism marketers and destination managers to understand the psychological and emotional aspects of travel, especially during this new normal, and, in particular, the capacity to create, shape and stimulate restorative experiences that enhance the psychological and social well-being of travellers. By learning more about emotional differences in such experiences, the tourism industry can gain valuable insights into how to improve the design and marketing of its products (Selstad, 2007). As a response to this pandemic, Australia is capable of rebuilding its domestic tourism industry. Tourism marketers should consider the influence of travel companionship on components of the service encounter by designing services and programs that cultivate favourable interactions with those accompanying visitors. This might include offering group packages and pricing options for friends to visit attractions or experience different activities that the destination offers. Many destinations offer family or couple packages, though options for friends are more limited. Group travellers have significantly more positive tourist experiences than solo travellers. Although interactions with companions may seem less controllable than interactions with service providers, providers can provide activities that allow mutual enjoyment and shared experiences, from group-based discounts and prices to selecting family-friendly communication channels and family-friendly programmes and services.

References

Ajzen, I. and Fishbein, M. (1980) *Understanding Attitudes and Predicting Social Behaviour*. Englewood Cliffs, NJ: Prentice-Hall.

Baumeister, R. and Leary, M. (1995) The need to belong: Desire for intepersonal attachment as a fundamental human motivation. *Psychological Bulletin* 117 (3), 497–529.

Beckmann, E., Ballantyne, R. and Packer, J. (1998) Targeted interpretation: Exploring relationships among visitors' motivations, activities, attitudes, information needs and preferences. *Journal of Tourism Studies* 9 (2), 14–25.

Brunner-Sperdin, A., Peters, M. and Strobl, A. (2012) It is all about the emotional state: Managing tourists' experiences *International Journal of Hospitality Management* 31 (1), 23–30. https://doi.org/10.1016/j.ijhm.2011.03.004.

Bunkše, E.V. (2012) Sensescapes: Or a paradigm shift from words and images to all human senses in creating feelings of home in landscapes. *Landscape Architecture and Art* 1 (1), 10–15.

Buunk, B.P. and Verhoeven, K. (2010) Companionship and support at work: A microanalysis of the stress-reducing features of social interaction. *Basic and Applied Social Psychology* 12 (3), 243–258.

Chamberlain, K. and Zika, S. (1992) Religiosity, meaning in life, and psychological well-being. In J.F. Schumaker (ed.) *Religion and Mental Health* (pp. 138–148). New York: Oxford University Press.

Chen, G., Huang, S. and Zhang, D. (2017) Understanding Chinese vacationers' perceived destination restorative qualities: Cross-cultural validation of the perceived destination restorative qualities scale. *Journal of Travel and Tourism Marketing* 14, 1–13.

Chiesura, A. (2004) The role of urban parks for the sustainable city. *Landscape and Urban Planning* 68 (1), 129–138. https://doi.org/10.1016/j.landurbplan.2003.08.003.

Changuklee, C. and Allen, L. (1999) Understanding individuals' attachment to selected destinations: An application of place attachment. *Tourism Analysis* 4 (3–4), 173–185.

Choo, H. and Petrick, J. (2015) The importance of travel companionship and we-intentions at tourism service encounters. *Journal of Quality Assurance in Hospitality & Tourism* 16 (1), 1–23.

Cimprich, B. (1993) Development of an intervention to restore attention in cancer patients *Cancer Nursing* 16 (2), 83–92.

Cohen, E. (1972) Toward a sociology of international tourism. *Social Research* 39 (1), 164182.

Coleman, D. and Iso-Ahola, S.E. (1993) Leisure and health: The role of social support and self-determination. *Journal of Leisure Research* 25 (2), 111–128.

Crompton, J.L. (1979) Motivations for pleasure vacation. *Annals of Tourism Research* 6 (4), 408–424

Crompton, J. (1981) Dimensions of the social group role in pleasure vacations *Annals of Tourism Research* 8 (4), 550–568.

Crossley, E. (2012) Affect and moral transformations in young volunteer tourists: Emotion in motion. In M. Robinson and D. Picard (eds) *Tourism, Affect and Transformation* (pp. 85–97). Abingdon: Routledge.

Curtin, S. (2009) Wildlife tourism: The intangible, psychological benefits of human-wildlife encounters. *Current Issues in Tourism* 12 (5–6), 451–474. https://doi.org/10.1080/13683500903042857.

Dann, G. and Jacobsen, J.K.S. (2003) Tourism smellscapes. *Tourism Geographies* 5 (1), 3–25.

De Bloom, J., Geurts, S.A.E., Taris, T.W., Sonnentag, S., de Weerth, C. and Kompier, M.A.J. (2010) Effects of vacation from work on health and well-being: Lots of fun, quickly gone. *Work and Stress: An International Journal of Work, Health, and Organisations* 24 (2), 196–216.

Deci, E.L. and Ryan, R.M. (2008) Hedonia, eudaimonia, and well-being: An introduction. *Journal of Happiness Studies* 9 (1), 1–11. https://doi.org/10.1007/s10902-006-9018-1.

Dillette, A.K., Douglas, A.C. and Andrzejewski, C. (2018) Yoga tourism – a catalyst for transformation? *Annals of Leisure Research*, 1–20. https://doi.org/10.1080/11745398.2018.1459195.

Falk, J.H., Reinhard, E.M., Vernon, C., Bronnenkant, K., Heimlich, J.E. and Deans, N.L. (2007) *Why Zoos & Aquariums Matter: Assessing the Impact of a Visit to a Zoo or Qquarium*. Silver Spring: Association of Zoos & Aquariums.

Filep, S. and Laing, J. (2018) Trends and directions in tourism and positive psychology. *Journal of Travel Research* 58 (3), 1–12

Filep, S., Laing, J. and Csikszentmihalyi, M. (2016) What is positive tourism? Why do we need it? In S. Filep, J. Laing and M. Csikszentmihalyi (eds) *Positive Tourism* (pp. 3–15). Abingdon: Routledge.

Filep, S., Macnaughton, J. and Glover, T. (2017) Tourism and gratitude: Valuing acts of kindness. *Annals of Tourism Research* 66, 26–36. https://doi.org/10.1016/j.annals.2017.05.015.

Filo, K. and Coghlan, A. (2016) Exploring the positive psychology domains of well-being activated through charity sport event experiences. *Event Management* 20 (2), 181–199. https://doi.org/10.3727/152599516X14610017108701.

Fiorillo, A. and Gorwood, P. (2020) The consequences of the COVID-19 pandemic on mental health and implications for clinical practice. *European Psychiatry* 63 (1), E32. https://doi.org/10.1192/j.eurpsy.2020.35.

Fredrickson, B.L. (2004) The broaden-and-build theory of positive emotions. *Philosophical Transactions of the Royal Society B: Biological Sciences* 359 (1449), 1367–1378. https://doi.org/10.1098/rstb.2004.1512.

Fredrickson, B.L., Tugade, M.M., Waugh, C.E. and Larkin, G.R. (2003) What good are positive emotions in crisis? A prospective study of resilience and emotions following the terrorist attacks on the United States on September 11th, 2001. *Journal of Personality and Social Psychology* 84 (2), 365–376. https://doi.org/10.1037/0022-3514.84.2.365.

Gascon, M., Triguero-Mas, M., Martinez, D., Dadvand, P., Forns, J., Plasència, A. and Nieuwenhuijsen, M.J. (2015) Mental health benefit of long-term exposure to residential green and blue spaces: A systematic review. *International Journal of Environmental Resident Public Health* 12, 4354–4379.

Gregory, A. and Fu, X. (2018) Examining family cohesion's influence on resort vacation satisfaction. *Journal of Hospitality and Tourism Insights* 1 (1), 54–64.

Hartig, T. (2011). Issues in restorative environments research: Matters of measurement. In B. Fernández-Ramírez, C. Hidalgo-Villodres, C.M. Salvador-Ferrer and M.J. Martos Méndez (eds) *Psicología ambiental 2011: Entre los estudios urbanos y el análisis de la sostenibilidad* [Environmental psychology 2011: Between urban studies and the analysis of sustainability]. Proceedings of the 11th Conference on Environmental Psychology in Spain). Almería, Spain: University of Almería & the Spanish Association of Environmental Psychology.

Hartig, T., Mang, M. and Evans, G.W. (1991) Restorative effects of natural environment experiences. *Environment and Behavior* 23 (1), 3–26. https://doi.org/10.1177/0013916591231001.

Hartig, T., Evans, G.W., Jamner, L.D., Davis, D.S. and Gärling, T. (2003) Tracking restoration in natural and urban field settings. *Journal of Environmental Psychology* 23 (2), 109–123.

Heung, V.C.S. and Kucukusta, D. (2013) Wellness in tourism in China: Resources, development and marketing. *International Journal of Tourism Research* 15, 346–359.

Huta, V. (2013) Pursuing eudaimonia versus hedonia: Distinctions, similarities, and relationships. In A.S. Waterman (ed.) *The Best within Us: Positive Psychology Perspectives on Eudaimonia* (pp. 139–158). Washington, DC: American Psychological Association.

Jensen, M.T., Scarles, C. and Cohen, S.A. (2015) A multisensory phenomenology of interrail mobilities. *Annals of Tourism Research* 53, 61–76.

Kaplan, S. (1995) The restorative benefits of nature: Toward an integrative framework. *Journal of Environmental Psychology* 15 (3), 169–182. https://doi.org/10.1016/0272-4944(95)90001-2.

Kim, S.S., Choi, S., Agrusa, J., Wang, K.C. and Kim, Y. (2010) The role of family decision makers in festival tourism. *International Journal of Hospitality Management* 29, 308–318.

Kim, H., Lee, S., Uysal, M., Kim, J. and Ahn, K. (2015) Nature-based tourism: Motivation and subjective well-being. *Journal of Travel and Tourism Marketing* 32 (1), 76–96. https://doi.org/10.1080/10548408.2014.997958.

Kirillova, K. and Wang, D. (2016) Smartphone (dis)connectedness and vacation recovery. *Annals of Tourism Research* 61, 157–169.

Korpela, K.M. and Ylén, M. (2007) Perceived health is associated with visiting natural favorite places in the vicinity. *Health & Place* 13 (1), 138–151.

Kottler, J.A. (1997) *Travel That Can Change Your Life: How to Create a Transformative Experience*. San Francisco: Jossey-Bass.

Lehto, X., Kirillova, K., Li, H. and Wu, W. (2017) A cross-cultural validation of the perceived destination restorative qualities scale: The Chinese perspective. *Asia Pacific Journal of Tourism Research* 2 (3), 329–343.

Leiper, N. (1979) The framework of tourism: Towards a definition of tourism, tourist, and the tourist industry. *Annals of Tourism Research* 6 (4), 390–407. https://doi.org/10.1016/0160-7383(79)90003-3.

Lin, C.-H. (2012) Effects of cuisine experience, psychological well-being and self health perception on the revisit intentions of hot springs tourists. *Journal of Travel Research* 38 (2), 243–265.

Ma, J., Gao, J., Scott, N. and Ding, P. (2013) Customer delight from theme park experiences: The antecedents of delight based on cognitive appraisal theory. *Annals of Tourism Research* 42, 359–381. https://doi.org/10.1016/j.annals.2013.02.018.

Matteucci, X., Volic, I. and Filep, S. (2019) Dimensions of friendship in shared travel experiences. *Leisure Sciences* 44 (6), 697–714. https://doi.org/10.1080/01490400.2019.1656121.

McCabe, M., Althof, S.E., Assalian, P., Measson M.-C., Leiblum, S.R., Simonelli, C. and Wylie, K. (2010) Psychological and interpersonal dimensions of sexual function and dysfunction *Journal of Sexual Medicine* 7 (1pt2), 327–336.

Miao, F.F., Vitterso, J., Ferssizidis, P., Fredrickson, B.L., Steger, M.F., Catalino, L.I. and Ryan, R.M. (2013) *Functional Well-Being: Happiness as Feelings, Evaluations, and Functioning*. Oxford: Oxford University Press.

Mitas, O., Qian, X.L., Yarnal, C. and Kerstetter, D. (2011) 'The fun begins now!': Broadening and building processes in Red Hat Society® participation. *Journal of Leisure Research* 43 (1), 30–55.

Nawijn, J. (2011) Happiness through vacationing: Just a temporary boost or long-term benefits? *Journal of Happiness Studies* 12, 651–665.

Nawijn, J., Marchand, M.A., Veenhoven, R. and Vingerhoets, A.J. (2010) Vacationers happier, but most not happier after a holiday. *Applied Research in Quality of Life* 5 (1), 35–47. https://doi.org/10.1007/s11482-009-9091-9.

Nawijn, J., Mitas, O., Lin, Y. and Kerstetter, D. (2012) How do we feel on vacation? A closer look at how emotions change over the course of a trip. *Journal of Travel Research* 52 (2), 265–274.

Ohe, Y., Ikei, H., Song, C. and Miyazaki, Y. (2017) Evaluating the relaxation effects of emerging forest-therapy tourism: A multidisciplinary approach. *Tourism Management* 62, 322–334.

Olsson, L.E., Gärling, T., Ettema, D., Friman, M. and Fujii, S. (2013) Happiness and satisfaction with work commute. *Social Indicators Research* 111 (1), 255–263.

Ottoson, J. and Grahn, P. (2005) A comparison of leisure time spent in a garden with leisure time spent indoors: On measures of restoration in residents in geriatric care. *Landscape Research* 30 (1), 23–55. https://doi.org/10.1080/0142639042000324758.

Packer, J. (2013) Visitors' restorative experiences in museum and botanic garden environments. In S. Filep and P. Pearce (eds) *Tourist Experience and Fulfilment: Insights from Positive Psychology* (pp. 203–210). New York: Routledge.

Pearce, P.L. (1987) Psychological studies of tourist behaviour and experience. *Australian Journal of Psychology* 39 (2), 173–182.

Pearce, P.L. (2009) The relationship between positive psychology and tourist behavior studies. *Tourism Analysis* 14 (1), 37–48. https://doi.org/10.3727/108354209788970153.

Pearce, P.L. (2012) Tourists' written reactions to poverty in Southern Africa. *Journal of Travel Research* 51 (2), 154–165. https://doi.org/10.1177/0047287510396098.

Pearce, P.L. and Lee Uk-II (2005) Developing the travel career approach to tourist motivation. *Journal of Travel Research* 43 (3), 226–237.

Pearce, P.L. and Packer, J. (2013) Minds on the move: New links from psychology to tourism. *Annals of Tourism Research* 40, 386–411. https://doi.org/10.1016/j.annals.2012.10.002.

Pyke, S., Hartwell, H., Blake, A. and Hemingway, A. (2016) Exploring well-being as a tourism product resource. *Tourism Management* 55, 94–105. https://doi.org/10.1016/j.tourman.2016.02.004.

Raymore, L.A. (2002) Facilitators to leisure. *Journal of Leisure Research* 34 (1), 37–51. https://doi.org/10.1080/00222216.2002.11949959.

Reitsamer, B. and Brunner-Sperdin, A. (2015) Tourist destination perception and well-being: What makes a destination attractive? *Journal of Vacation Marketing* 23 (1), 55–72.

Rook, D.W. (1987) The buying impulse. *Journal of Consumer Research* 14 (2), 189–199. https://doi.org/10.1086/209105.

Rojas-De-Gracia, M.M. and Urbistondo, P. (2019) Couple's decision-making process and their satisfaction with the tourist destination. *Journal of Travel Research* 58, 824–836. https://doi.org/10.1177/0047287518785052.

Rosen, R.S. (2018) Geographies in the American DeafWorld as institutional constructions of the deaf body in space: The sensescape model. *Disability & Society* 33 (1), 59–77.

Seligman, M.E. (2004) *Authentic Happiness: Using the New Positive Psychology to Realize Your Potential for Lasting Fulfillment*. New York: Simon & Schuster.

Selstad, L. (2007) The social anthropology of the tourist experience. Exploring the 'Middle Role'. *Scandinavian Journal of Hospitality and Tourism* 7 (1), 19–333.

Sharpley, R. and Jepson, D. (2011) Rural tourism: A spiritual experience? *Annals of Tourism Research* 38 (1), 52–71. https://doi.org/10.1016/j.annals.2010.05.002.

Sirgy, M.J. and Lee, D.-J. (2008) Well-being marketing: An ethical business philosophy for consumer goods Firms. *Journal of Business Ethics* 77 (4), 377–403. http://www.jstor.org/stable/25075572.

Sirgy, M.J. and Uysal, M. (2016) Developing a eudaimonia research agenda in travel and tourism. In M.J. Sirgy and M. Uysal (eds) *Handbook of Eudaimonic Well-Being* (pp. 485–495). Cham: Springer.

Smith, M.K. and Diekmann, A. (2017) Tourism and wellbeing. *Annals of Tourism Research* 66, 1–13. https://doi.org/10.1016/j.annals.2017.05.006.

Tugade, M.M. and Fredrickson, B.L. (2004) Resilient individuals use positive emotions to bounce back from negative emotional experiences. *Journal of Personality and Social Psychology* 86 (2), 320–333. https://doi.org/10.1037/0022-3514.86.2.320.

Ulrich, R.S., Simons, R.F., Losito, B.D., Fiorito, E., Miles, M.A. and Zelson, M. (1991) Stress recovery during exposure to natural and urban environments. *Journal of Environmental Psychology* 11 (3), 201 230.

Um, S. and Crompton, J.L. (1990) Attitude determinants in tourism destination choice. *Annals of Tourism Research* 17 (3), 432–448. https://doi.org/10.1016/0160-7383(90)90008-F.

Uysal, M., Perdue, R. and Sirgy, M.J. (eds) (2012) *Handbook of Tourism and Quality-of-Life Research: Enhancing The Lives of Tourists and Residents of Host Communities*. Dordecht: Springer.

Uysal, M.M., Sirgy, N.J., Woo, E. and Kim, H. (2015) Quality of life (QOL) and well-being research in tourism *Tourism Management* 53, 244–261.

Vada, S., Prentice, C. and Hsiao, A. (2019) The influence of tourism experience and well-being on place attachment. *Journal of Retailing and Consumer Services* 47 (C), 322–330.

Vada, S., Scott, N., Prentice, C. and Hsiao, A. (2020) Positive psychology and tourist well-being: A systematic literature review. *Tourism Management Perspectives* 33, 1–14.

Van Den Berg, A., Hartig, T. and Staats, H. (2007) Preference for nature in urbanized societies: Stress, restoration and the pursuit of sustainability. *Journal of Social Issues* 63 (1), 79–96.

Voigt, C., Howat, G. and Brown, G. (2010) Hedonic and eudaimonic experiences among wellness tourists: An exploratory study. *Annals of Leisure Research* 1 (13), 541–562. https://doi.org/10.1080/11745398.2010.9686862.

World Travel and Tourism Council (WTTC) (2019) Economic impact reports. London: WTTC.

Zhu, J. and Fan, Y. (2018) Daily travel behaviour and emotional well-being: Effects of trip mode, duration, purpose, and companionship. *Transportation Research Part A: Policy and Practice* 118, 360–373.

12 Tourism, Hope and Peace: A Counter-Discourse in Palestine

Rami K. Isaac

Introduction

On 17 December 2014, a vast number of Members of the European Parliament voted to recognise the state of Palestine. The motion stipulated that the European Parliament supports 'in principle the recognition of Palestinian statehood and the two-state solution and believes these should go hand in hand with the development of peace talks, which should be advanced' (Europarl, 2014). The Israeli stance on such advancements was reflected in the then Israeli Prime Minister Benyamin Netanyahu's response that any attempts at forcing such conditions on Israel can only lead to a deterioration of the political situation in the region and endanger the State of Israel (RT, 2014).

According to Isaac *et al.* (2016: 1), tourism in Palestine has been gaining an increasingly important profile given its economic and religious significance, as well as the substantial role it can play in Israeli–Palestinian relations vis-à-vis representations of Palestinian statehood and identity. On the one hand, the Palestinian reality and daily life under occupation has been widely erased from the Israeli-controlled tourism sector in Palestine (Van den Boer, 2016). On the other hand, many visitors who are drawn to the region, either due to their religious and ethnic affiliations or due to their political ideologies, feel connected to the conflict and wish to better understand it. As a response to this demand of the market, several non-governmental organisations (NGOs) and peace and social organisations have begun to emerge in Palestine, to showcase their work and increase their support (Brin, 2006; Isaac, 2010a). In the context of the ongoing political conflict, some of these organisations have engaged tourism as an important instrument in (re)shaping and rearticulating the imaginative geographies related to Palestine.

During the 21st century, several scholars have shown interest in the role of tourism as a social force (Higgins-Desbiolles, 2006b) that might be able to promote tolerance and international understanding among cultures (Farmaki, 2017). According to Farmaki (2017), as such, the contextual issues surrounding the development of tourism in past and current conflict-ridden locations tend to be strongly linked to the degree to which tourism can contribute to peace. Farmaki (2017: 529) states, '[T]he links between tourism and peace raise some important questions, such as: What type of tourism is appropriate for peacebuilding? [A]nd what forms of peace does tourism contribute to?'. There is also the important question of *how* or *what* exactly tourism may contribute to peace. It is this question that forms the focus of this chapter, by exploring how tourism can play a role as peacemaker in the case of Palestine and Israel. The author focuses on forms of alternative tourism that offer a counter-discourse in the region, and, in particular, how these examples of alternative tourism help to create a counter-discourse through the construction of narratives of hope (Tucker & Shelton, 2018). In this emergent counter-discourse, the main intention is to give voice to the Palestinian lifeworld in order to counteract the Israeli dominant discourse. As part of a counter-discourse, these alternative tourism organisations support the resistance against the dominant Israeli discourse.

The Context of Palestine and Israel

In the state of Palestine, the government is represented by Israel and the Palestinian Authority subordinated to Israel. Moreover, the context of Palestine and Israel is unique due to the physical divide created by the ongoing conflict between both countries. As such, the Israeli–Palestinian conflict has severely impacted the tourism sector. The lack of sovereignty of the Palestinian government over its territories, on the one hand, and the fragmentation of the land of Palestine between the West Bank, the Gaza Strip and East Jerusalem, on the other, has affected visitor access and mobility into and around Palestine (Isaac & Platenkamp, 2012). Moreover, as Van den Boer (2016: 1) states, '[T]he Zionist settler colonial project has rendered Palestinians invisible, not just by forcing them behind walls but also by co-opting their food, crafts and folklore or representing them as the eternal "other"'. In this context, tourism is a practice that normalises these images and reorders and reshapes perceptions of people, places and their relationship to the world (Hollinshead, 2009).

Currently, Palestinians still do not have control over their economy. The Oslo agreement with Israel installed a system of fragmentation that consolidated and deepened Israeli control over all aspects of Palestinian life, rendering a Palestinian state unviable (Tabar & Salamanca, 2015). Lisle (2000: 93) argues that entanglements between conflict and tourism 'disrupt and resist the prevailing images of safety and danger that attempt to hold them apart'. Studying these complex interconnections 'prevents the hegemonic discourses of global security from completing itself, stabilizing its boundaries and securing a totalized presence' (Lisle, 2000: 93).

These hegemonic discourses create a separation between sociopolitical conflicts and tourism through frequent reminders of the importance of safety and security in tourism.

The Palestinian tourism industry has consequently flourished with increasing numbers of visitors despite the many difficulties that come in the shape of checkpoints, closures and lack of border control (Isaac, 2018). Indeed, Buda (2015) has written in-depth about the 'attraction' of these checkpoints. Buda's argument is that the Israeli–Palestinian conflict provides a context for some 'adventure tourists' to engage in a ludic form of death drive, which in turn produces a sense of aliveness for these adventure-seeking tourists (2015: 134).

In addition to 'death drive' tourism, another 'alternative' type of tourism is developing that caters to tourists seeking to understand life under occupation in Palestine (White, 2015). The definition of and correct terminology for alternative tourism are contentious (Higgins-Desbiolles, 2008). Alternative tourism is still a vague concept that lacks concrete definition (Butler, 1992: 31; Pearce, 1992: 15). There are many types of alternative tourism, such as soft tourism, special interest tourism, and many others. These are all united in that they are opposed to mass tourism. However, alternative tourism in Palestine has an innovative approach, presenting a critical look at the culture, history and politics of Palestine and its complex relationship with Israel (Isaac, 2010a). These types of organisations, such as the Alternative Tourism Group (ATG), '[operate] according to the views of justice tourism, that is, tourism that holds as its central goals the creation of economic opportunities for the Palestinian community, positive cross-cultural exchange between guest and host through one-on-one interaction' (Isaac, 2010a: 30). Justice tourism (Isaac & Hodge, 2011) in the context of Palestine involves forms of tourism that allow travellers to encounter the 'truth' of Israeli oppression and Palestinian suffering and to offer a chance to act in solidarity with the Palestinian people. These forms of tourism could potentially contribute to the production of a counter-discourse to the dominant Israeli one and hence constitute a form of 'alternative tourism'. Many studies have been conducted on these forms of (alternative) tourism, such as political-oriented tourism (Clarke, 2000; Brin, 2006; Moynagh, 2008), solidarity tourism (Bowman, 1992; Isaac, 2010a; Kassis *et al.*, 2016; Noy, 2011), justice tourism (Isaac & Hodge, 2011; Higgins-Desbiolles, 2008) and activism (Belhassen *et al.*, 2014). In this context, tourism creates realities that people consider their own. Realities are also representations and reflections that do not stand isolated but are produced in relation to the wider socioeconomic and political context in which Palestinians live.

While this 'alternative tourism' continues to represent only a minor segment of the tourism industry in Palestine, it is nevertheless increasing and is hence worthy of examination, particularly in relation to how, or what, it may potentially contribute to peace and hope in the region. That, indeed, is the focus of the present chapter, which is premised on the following research question: What, or how, might 'alternative tourism', through construction of a 'hopeful' narrative, contribute to peace in the context of Palestine and Israel?

Tourism and Peace

Several scholars (Butler & Mao, 1995; Selwyn & Karkut, 2007; Yu, 1997) have criticised the proposition that tourism has potential to reduce conflicts between divided countries, while other researchers have considered that tourism may act as a positive instrument through reducing tension and suspicion (Hobson & Ko, 1994; Richter, 1994; Var et al., 1989). Until today, this suggestion has not always been documented, with Litvin (1998), for example, suggesting that tourism is not a generator of peace but the beneficiary of it.

Over the years, political tourism has sprung up in former and ongoing conflict-ridden destinations such as South Africa (Van Amerom & Buscher, 2005), North and South Korea and Cyprus (Timothy et al., 2004). Several scholars have hypothesised that contact between peoples through travel heightens tourism's role as an agent of change, which may bring down obstacles between people and encourage cooperation between nations (Sarkar & George, 2010). Others have examined the role of tourism as an instrument for peacebuilding (Alluri, 2009; Causevic, 2010; Causevic & Lynch, 2011; Durko & Petrick, 2016; Guo et al., 2006). All these studies have been conducted on the proposition that tourism can be an agent for peace in establishing positive relationships between separated communities.

Kim and Crompton (1990) introduced the concept of two-track diplomacy. The first track is described as official government-to-government relations while the second is an unofficial means of people-to-people relations through tourism. Track Two is a grassroots type of engagement that campaigners of justice tourism believe to be the ideal way to foster understanding between people (Isaac & Hodge, 2011).

While it is apparent that the relationship between peace and tourism remains unclear, the relationship between the various manifestations of alternative tourism and peace is arguably most poignant in their ability to promote critical and reflective thinking (Inui et al., 2006). Foucault, whose concept of a discourse has been mainly introduced into tourism discussions by Urry (1990), states that a discourse is a way of constituting knowledge, together with the social practices, forms of subjectivity and power relations that are inherent in such knowledge and relations between them. Through such discourse, tourism has been used to erase Palestinian presence and narratives from Palestine. The process of disciplining (Foucault, 1975) and exclusion, in this case of the Palestinian people, are part of the mechanism that Foucault refers to in his concept of a discourse. According to Foucault, counter-discourses emerge at the same time despite the dominant discourse. In relation to the situation in Palestine, there are dispersed elements of such a counter-discourse that are produced through alternative forms of tourism. This also refers to the necessity of human agency; here human agency becomes an element underestimated by the Foucauldian power–knowledge constellation related to the concept of (counter)discourse. As Arendt (1958: 247) posited, 'The miracle that saves the world, the realism of human affairs, from its normal, "natural" ruin is ultimately the fact of natality, in which the faculty of action is ontologically rooted'. Through Habermas's (1984) communicative action it becomes possible to support this necessary role of 'human agency'. As Isaac and Platenkamp (2016: 159) assert, '[C]ommunicative action never stops from within life-worlds, and human agency is embedded in it. Whereas the Israeli discourse colonises the Palestinian lifeworld and excludes deviant opinions because of it, communicative action within this same lifeworld can organize the dispersed elements of a counter-discourse to be revitalized.' In this emergent counter-discourse, the main intention is to give voice to the Palestinian lifeworld in order to counteract the devastating mechanism of 'obscuring the real condition of society' (Mannheim, 1936: 36).

In this context, tourism is a part of a hegemonic colonial project but, in addition to that, tourism can be a potential site of anticolonial struggle and unsettling hegemonic knowledge, representations, practices and realities as advocated elsewhere (Grimwood *et al.*, 2019). Tourism is not a neutral activity of 'innocents abroad' as Mark Twain famously labelled it (Twain, 2007).

In the context of Palestine, 'special interest tourism' may include forms of 'war tourism', 'conflict tourism' or other politically-oriented forms of tourism. Politically-oriented tourists (Brin, 2006), conflict tourists (Warner, 1999), danger-zoners (Buda, 2015), dark tourists (Isaac & Ashworth, 2012) and war tourists (Pitts, 1996) are all considered types of dark tourists who travel 'to places made interesting for reasons of political dispute' (Warner, 1999: 137). War tourists, as argued by Pitts (1996: 224), are the ones for whom political conflict represents the principal reason for travelling in a region, and their main motivation is 'to experience the thrill of political violence'. Danger-zoners, according to Adams (2001: 266), show interest in the ongoing political clashes of a region and experience ongoing conflicts first-hand. As already mentioned, danger-zone tourism, defined as tourism that thrives in tumultuous times, is discussed by Buda (2015) in relation to the Israeli–Palestinian context and is associated very much with the concept of 'dark tourism'.

However, continuing to depict tourism in this context as dark tourism could be considered unhelpful with regards to thinking about the role that tourism might play, both in creating counter-narratives to conflict and in creating narratives of hope in order to 'counter' the sense of despair. On a political level, several authors have examined tourism as a method of peaceful resistance against the Israeli oppression and argue that this kind of tourism can help raise awareness on the Palestinian cause (Griffiths, 2016; Isaac, 2010a, 2010b; Isaac *et al.*, 2016). This form of alternative tourism can therefore be seen as contributing towards a narrative of hope. Based on a general understanding that hope matters (Ojala, 2012), the ways in which tourism might become one of the avenues for instilling hope is discussed in-depth by Tucker and Shelton (2018). In particular, they examine the ways in which narrative constructions in tourism contexts – for example, the narratives constructed and presented by guides in guided tours – might include stories that explicitly anticipate something better in the future, thereby producing a hopeful counter-discourse to the dominant (Israeli) political discourse. Indeed, if this is so, then such narrative constructions in the context of alternative tours in Palestine context might be considered anything but dark.

In an unequal power context such as Palestine and Israel, the question thus remains as to what role tourism can play in peacebuilding initiatives and in helping to strengthen not only a counter-discourse but also a hopeful one, especially while any prospect of reaching a diplomatic resolution in the foreseeable future seems dim.

Methodology

This research is an exercise in tourism geography as it seeks to understand tourism's relationship with the environment within which it takes place (Mitchell & Murphy, 1991). It takes an inductive approach whereby we have allowed direct field experience to inform theory development (Patton, 2002). Given that this study explores the counter-narratives of space, we used the method of interpreting narratives – that is, qualitative methods. We have combined two core elements of qualitative work: participant observations and in-depth interviews. Fieldwork, two months in total, was conducted in Palestine in 2018 and 2019. During these visits, extensive participant observation was conducted where the focus was on understanding the tours offered by selected organisations, particularly the ATG and Kairos Palestine. This study interrogates the examples of 'alternative tours' to see how a tourism tour tries to challenge the existing dominant narratives that characterise the Israeli–Palestinian space. As part of this participant observation, notes were continually written on how Palestinian guides interpret the narratives and stories for visitors, for example regarding the Segregation Wall around Bethlehem, the refugee camps, and the city of Hebron. In-depth interviews were conducted with selected key persons chosen based on their knowledge and involvement, including the director of the ATG, the director of Kairos Palestine, and two Palestinian tour guides. They shared their experiences around the topic of the role of tourism in building peace, equality and human rights for all the people in the country. All interviews were transcribed and the data were then coded and analysed for explanatory themes, which have informed the way we have structured the findings.

Introducing Alternative Tours in the Palestinian–Israeli Context

The Alternative Tourism Group

The ATG is a Palestinian NGO organisation, founded in 1995 and located in the town of Beitsahour in the governorate of Bethlehem. The ATG presents a critical look at the history, culture, and politics of Palestine and its complex relationship with Israel and the Israeli occupation of Palestine. According to the director (Kassis), the ATG operates according to the tenets of justice tourism and equality – that is, tourism that holds its central aim to be the creation of economic opportunities for the Palestinian local community and positive understanding between hosts and guests through face-to-face interaction. The ATG was founded when many Palestinians felt that their contemporary culture and the political realities they were living did not find adequate expression in conventional pilgrim tourism (the mainstream form of tourism in Bethlehem).

Kairos Palestine: 'Come and see'

Kairos Palestine is a Christian movement group that advocates ending the Israeli occupation. Kairos Palestine co-authored a document called 'A Moment of Truth: A Word of Faith, Hope and Love from the Heart of the Palestinian Suffering', which is Christian Palestinians' call for international solidarity in establishing a just peace in the region (Kassis, 2011: 21). It is in the name of these goals that Kairos Palestine has both spearheaded and participated in a wide range of advocacy initiatives, workshops, conferences and publications on local, national, regional and international levels – all seeking not only to bolster and prioritise the Palestinian struggle for peace with justice but also to emphasise peace with justice (Kassis, 2011). One of their initiatives in relation to tourism is the campaign entitled 'Come and see', a call from Palestinian Christians to a journey for peace with justice (Isaac, 2021).

These two organisations create a new reality on the ground and position Palestinians as a native population that is disposed of, excluded and eliminated while at the same time including stories of hope for the future, thereby producing a 'hopeful' counter-discourse to the dominant (Israeli) political discourse.

Findings

Two major themes emerged in relation to the ways in which these tours counter the Israeli–Palestinian context that these organisations aim to challenge. The first is how they tell the story of occupied Palestine to international tourists, thereby presenting a hopeful narrative, and the second is how these tours challenge the dominant Israeli narrative by demonstrating that Palestinians can create opportunities and spaces for increased engagement between Israelis and Palestinians.

Stories through alternative tours, a hopeful narrative for a better future

One Palestinian tour guide (labelled Tour Guide A) stated:

I take visitors to the 'places of bother' such as the Segregation Wall at the entrance of Bethlehem. I first tell them the story of the construction of the wall and the consequences of the wall on the livelihood of the area, which used to be one of the commercial hubs in Bethlehem. Then I explain that, on that exact point, earlier, people were living here, with shops, restaurants and [it] used to be a lively city, but now it is a complete, a ghost area. Tourists usually remain silent to reflect on the tremendous force of this wall. There were so many houses, and people living their daily lives.

Such a situation underscores what might be referred to as a communication action of affect, which often resides outside the domain of language, but which is still visible to embodied practices (Richard & Rudnyckyj, 2009).

One common practice that tour Palestinian tour guides 'utilise to make tourists "understand" the sociopolitical conflict and the consequences of occupation is . . . showing them pictures of the area before the construction of the Segregation Wall, often with only brief explanations in order to let tourists feel the magnitude of what happened here' (Tour Guide A). The implications of these activities and life under occupation do provide tourists counter-discourses about the daily life of the Palestinians, which could eventually provide a sense of hope for Palestinians despite oppression by the Israeli occupation.

In addition, Palestinian guides usually leave some time for tourists to just remain in silence and confront the experience. Another Palestinian tour guide (labelled Tour Guide B) shared his feelings about the tour and explained that 'silence feels for many tourists the only appropriate response when the scale of the experience exceeds any narration, and witnessing in which the power of silence acts [as] a chamber of resonance sense of hope'.

Visits to these tangible landmarks, such as the 26-foot-high concrete wall, corroborated by narratives and onsite practices regularly repeated with every new group of tourists, are aimed at creating a structure directed by the tour guide (see Edensor & Holloway, 2008) that produces and reproduces a specific result of the occupation landscape. In the interviews we conducted, it was clearly stated by tour guides and alternative tour organisations that 'the local population understands international tourists as a key source for creating a sense of hope, but also economic revenue for the area'. Tour Guide B in addition commented that 'the objective of these encountersfor tourist [is] to break down the negative stereotypes of Palestine and its people that predominate in the Western countries'.

The director of the ATG put it this way: 'International tourists are genuinely conceived as an opportunity for residents and refugees to tell their stories and have their story known outside of Palestine. Tourists were expected to go back home and communicate, on their own terms'. He continued, 'These tourists were not only regarded as witnesses but also as participants in the realisation of a new, hopeful bright future'. It becomes clear how these alternative tourism initiatives work to reorder reality and that through them the oppressed can make use of tourism to both circulate their claims and make them solid, thereby presenting a hopeful narrative.

The director further stated that 'ATG seeks to promote a positive image of Palestine and its people and to contribute towards establishing rightful peace in the area. When Palestinian[s] behave as self-conscious citizens, who are aware of their oppression and domination and trying to create hope for a better future, these tourists can partake in this process'.

In the experience of the director of ATG, 'through these reactions of tourists, who share the feelings of sadness of the Palestinian people, the process of creating hope becomes a crucial topic for these tourists. The creation of hope that is involved in this process and which needs a place for development in this overwhelming circumstance, this finds a "hopeful" niche in alternative forms of tourism'.

Taking international tourists to 'places of bother' (Isaac & Platenkamp, 2012) as represented by landmarks, narratives, and performances by the tour guide – or the very people (refugees) who have experienced expulsion and displacement from their hometowns – can greatly benefit tourists, who can go back home and tell their stories with a sense of understanding and hope, emphasised with and hopeful for something better in the future. As the director of Kairos Palestine put it, 'In the end, such visitors or pilgrimages "become holders of the knowledge that will one day lead to equality, democracy, and human rights for all"'. This is clearly supporting the communicative action that is needed in the creation of hope.

Giving Palestinians a voice to counteract the Israeli dominant discourse

The main idea behind ATG's tours was outlined by AYG in our interview: 'Our founding principle is the establishment of just and responsible tourism where visitors are engaged and participate to gain a better understanding of the political reality on the ground. The goal is to raise awareness amongst visitors and allow them to become witnesses and advocates for peace and justice.' The overall objective of these tours is to expose tourists to counter-narratives (discourse) about Israel in the context of ongoing conflict. They promote the visibility of Palestinians, exposing tourists to Palestinian daily life under occupation in Palestinian neighbourhoods.

Talking to Tour Guide A during a tour in Hebron, where around 80% of the city population are Palestinians (known as H1) and around 20% are Jewish (living in the area containing the settlements and known as H2, which is guarded by the Israeli military occupation), he made it clear that resistance was part of the reasons for engaging in these tours. By going to places such Hebron that are threatened by Israeli settlers and face isolation due to the Segregation Wall, these tours aim to reconnect to these spaces, demonstrating their presence as to reclaim them as essential part of Palestine.

Once more, these experiences reshape how people and tourists perceive Palestine as the Palestinian homeland and this is challenging the way the Israeli colonial power is trying to disprove connections between the Palestinian population and the land of the West Bank in general and Hebron in particular. These tours are aiming to produce both space and knowledge in a way that confronts the hegemonic settler logic.

The executive director of ATG, Kassis, stated:

> ATG [has] chosen to create innovative tourist packages of [an] all-encompassing nature, including interpretive tours, homestays, and cultural experiences and visits to mainstream sites. These would serve the basic requirements of a 'normal' tourist who visits Palestine. What makes it different, however, and crucial for the counter-discourse, is the narrative, which makes tourism such a powerful tool to reshape and rearticulate places, identities, people, particularly in Palestine. Tourist[s] are being introduced to various structures of occupation, the movement restrictions of Palestinians and the daily life of Palestinian people.

Tour Guide B shared that 'the destruction caused by the Segregation Wall, for example, can allow tourists access to a truth, which triggers an emotional involvement that opens the door to an internal process through which truth is exposed, and something unexpectedly is understood without words and explanations'.

This racialised system of mobility has taken different shapes over the years, which include checkpoints, bypass roads, the Segregation Wall, closure, and curfews. Interestingly here, too, tour guides aim to take tourists into a different world, one that gives space to the Palestinian narrative and challenges the colonial fabrications that dominate much of the mass tourism coming to Israel and Palestine. In this vein, Palestine is coproduced through various touristic interventions and commodities such as maps, guidebooks, and other things valued by tourists which now become representations of what is valued by Palestinians.

According to the director of Kairos Palestine:

[T]he crucial importance of Kairos Palestine is its call to 'Come and See'. Many tourists and visitors come to Palestine and Israel as visitors, but they do not see. Kairos emphasises the power of 'seeing' because they believe that for many people of goodwill – as is true of the many tourists and visitors who come to Palestine and Israel – a clear vision of the reality around them is enough for them to be transformed.

The director of Kairos Palestine continued, saying that the aim of this organisation is 'requesting the international community to stand by the Palestinian people who have faced oppression, displacement, and clear apartheid for more than six decades. The suffering continues while the international community silently looks at the occupying State of Israel.' People-to-people interactions, in addition, have become significant opportunities for increasing engagement between Israelis and Palestinians.

According to Tour Guide B, 'The information and analysis provided on the tours in collaboration with Green Olive Tours in Israel are based on human rights reflections, the rights of self-determination, and the rights of all peoples to live in safety and freedom'.

Palestinian Tour Guide A described how 'story-telling in Bethlehem covers basic facts but, most importantly, it includes also personal stories told by the guides and affected people, for example, the house of Anastas's family in Bethlehem. . . . [T]he Segregation Wall surround[s] her house on three sides. . . . [T]ourists stay silence (sic) experiencing this house'.

As noted by Willis (2014: 6), the 'complex role of presence and spectatorship [is] often related to a sense of being at loss, not so much [because of the] sadness that comes from seeing something profoundly moving, but rather [because of] the unease of not knowing how to respond'. The set-up of the tours also allows tourists to interact with Anastas's family, to get to know the diverse communities living near the Segregation Wall and to experience the complexities of their daily lives, rooted in politics.

As the director of the ATG states, 'Tourism does have a substantial part to play in developing peaceful relations between these two divided communities and can positively influence current international visitors on "both sides" in forwarding the message to their friends and relatives'.

Through this increasing engagement, alternative tours can contribute to the peace process within the frame of this emerging counter-discourse. People-to-people contacts created by trade and services (tourism), such as by the Palestinian and the Israeli tour operators, can lead to an increase of wealth and the creation of a middle class who can articulate the desire and hope for political freedoms such as democracy and human rights.

Discussion

Tourism in the context of this study opens a new space to investigate asymmetric power relations and oppression. Several authors have encouraged a rethinking beyond tourism's hegemonic embeddedness and see it as a potential force of emancipation (e.g. Higgins-Desbiolles, 2006a). It can be argued that once the visitor is aware of the nature of the interaction between Israelis and Palestinians, they might be more empathetic to the Palestinian people, who are still under occupation. This is indeed a part of the emerging counter-discourse (see also Shepherd & Laven, 2019). This kind of alternative tourism has a long legacy in Palestine. During and after the First Intifada which began in December 1987, many foreigners, particularly church groups and other fact-findings groups, came to Palestine in solidarity. In this regard, it borrows some of the premises of justice tourism as defined by Scheyvens (2002: 104), where visitors can partake in 'the liberation process', and by Isaac and Hodge (2011: 101), who see it as having the potential to resolve 'political issues and raising the awareness about the dramatic situation in Palestine'. In this way, tourism contributes to decolonised knowledge production. These new forms of tourism can therefore help to present alternative viewpoints externally and to trouble the dominant narratives pertaining to the Israeli–Palestinian conflict. They thereby also trigger empathy (Tucker, 2016) in politically oriented tourists, as well as in solidarity groups' home audiences, and they also contribute to the creation of hope. This demonstrates the appealing emergence of this kind of counter-discourse as a way of offering resistance and therefore a sense of hope.

Consequently, a much-needed platform could be developed via tourism activity in Palestine in order to create this little bit of hope. Broadly speaking, hope is linked to agency. As McGeer (2004: 105) argues, to hope 'is to experience ourselves as agents of potential as well as agents in fact'. In this way, hope is also linked to the notion of social or political change (Drahos, 2004; Freire, 1994; Ojala, 2012), often being seen as 'the fuel for agency' (Lueck, 2007: 256). Drahos (2004: 22) argues that hope 'carries the individual forward to the time of the hoped-for outcomes' in that it 'leads into a cycle of expectation, planning, and action'. Therefore, these different forms of alternative tourism are proactively challenging the narrative discourse and at the same time creating hope and thereby the fuel for agency, action, and change. In recent years, several authors who have shown interest in hope theory have also considered the connection between hope and agency, or action (Drahos, 2004; Freire, 1994; Ojala, 2012), while others have signalled that hope can lead to 'a negation or deferral of life' (Zournazi, 2002: 151). As Tucker and Shelton (2018: 72) suggest, it may promise happiness in some 'other' place or some 'other' time.

The illustrations of tourism narratives in Palestine show that the performance of hopeful narratives could create a hopeful mood in visitors. We can perhaps hope that tourists will go back home and tell their friends and relatives that a hopeful mood produced through tourism may prompt some sort of 'opening up of the future and opening up of what is possible' (Ahmed, 2004: 171).

As Tucker and Shelton (2018: 73) state, 'If a hopeful mood can fuel agency and action, and a pessimistic mood may lead to apathy, then it is clearly important to think about affect and moods produced through tourism in respect to what they might do vis-à-vis our being in, and our engaging with, the world'.

Today, in the geography of Israel and the occupied territories of Palestine, there is rough parity in numbers between Israeli Jews and Palestinians, but more than half of the Palestinian people are denied sovereignty over their land, full benefits of citizenship in a state, and equal rights. Therefore, the 'alternative tours' described here represent 'voices of hope' that echo that of the broader Palestinian society.

Conclusion

This chapter has aimed to explore the potential role of tourism to create peace and understanding through social relations that can be fostered with alternative tourism. In particular, the discussion has outlined alternative tours in Palestine, which can contribute to a counter-discourse and thereby produce a hopeful narrative. The overall objective of these tours, such as those by the ATG and Kairos Palestine, is to expose tourists to counter-narratives about Israel in the context of ongoing conflict. They promote the visibility of Palestinian narratives, exposing tourists to Palestinian life under military occupation in various territories in Palestine. Two major themes emerged in relation to the ways in which these tours counter the Israeli–Palestinian context that these organisations aim to challenge. The first is how they tell the story of occupied Palestine to international tourists, thereby presenting a hopeful narrative, and the second is how these tours are challenging the dominant Israeli narrative by demonstrating that Palestinians create opportunities and spaces for increased engagement between Israelis and Palestinians.

The empirical data presented by these tours have demonstrated how tourism is a productive process that becomes intertwined into influential recollections that produce much more than just a touristic experience. The findings show that tourism does give people a sense of hope. This chapter has demonstrated how alternative tours can facilitate the rearticulating and reordering of spaces to help to create a counter-discourse through the construction of narratives of hope.

References

Adams, K.M. (2001) Danger-zone tourism: Prospects and problems for tourism in tumultuous times. In T.C. Teo and K.C. Chang (eds) *Interconnected Worlds: Tourism in Southeast Asia* (pp. 265–281). Oxford: Pergamon.

Ahmed, S. (2004) *The Cultural Politics of Emotion*. Edinburgh: Edinburgh University Press.

Alluri, R.M. (2009) *The Role of Tourism in Post-Conflict Peace-Building in Rwanda*. Bern: Swisspeace.

Arendt, H. (1958) *The Human Condition*. Chicago: The University of Chicago Press.

Belhassen, Y., Uriely, N. and Assor, O. (2014) The touristification of a conflict zone: The case of Bil'in. *Annals of Tourism Research* 49, 174–189.

Bowman, G. (1992) The politics of tour guides: Israeli and Palestinian guides in Israel and occupied territories. In D. Harrison (ed.) *Tourism and the Less Developed Countries* (pp. 121–134). London: Harsted Press.

Brin, E. (2006) Political-oriented tourism in Jerusalem. *Tourist Studies* 6 (3), 215–243.

Buda, D.M. (2015) The death drive in tourism studies. *Annals of Tourism Research* 50, 39–51.

Butler, R. (1992) Alternative tourism: The thin end of the wedge. In W.R. Eadington and V.L. Smith (eds) *Tourism Alternatives: Potentials and Problems in the Development of Tourism*. Philadelphia: University of Pennsylvania Press.

Butler, R. and Mao, B. (1995) Tourism between divided quasi-states: International, domestic or what? In R. Butler and D. Pearce (eds) *Change in Tourism: People, Places, Processes* (pp. 92–113). London: Routledge.

Causevic, S. (2010) Tourism which erases borders: An introspection into Bosnia and Herzegovina. In I. Kelly and O. Moufakkir (eds) *Tourism, Progress and Peace* (pp. 48–64). Wallingford: CABI.

Causevic, S. and Lynch, P. (2011) Phoenix tourism: Post-conflict tourism role. *Annals of Tourism Research* 38 (3), 780–800.

Clarke, R. (2000) Self-presentation in a contested city: Palestinian and Israeli political tourism in Hebron. *Anthropology Today* 16 (5), 61–85.

Drahos, P. (2004) Trading in public hope. *The Annals of the American Academy* 592, 18–38.

Durko, A. and Petrick, J. (2016) The Nutella project: An education initiative to suggest tourism as a means to peace between the United States and Afghanistan. *Journal of Travel Research* 55 (8), 1081–1093.

Edensor, T. and Holloway, J. (2008) Rhythmanalysing the coach tour: The Ring of Kerry, Ireland. *Transactions of the Institute of British Geographers* 33 (4), 483–501.

Europarl (2014) New European Parliament resolution on recognizing of Palestine statehood. See https://europarl.europa.eu/news/en/press-room/20141212ipr01105/european-parliament-resolution-on-recognition-of-palestine-statehood (accessed June 2021).

Farmaki, A. (2017) The tourism and peace nexus. *Tourism Management* 59, 528–540.

Foucault, M. (1975) *Surveiller et punir*. Paris: Gallimard.

Freire, P. (1994) *Pedagogy of Hope*. New York: Continuum.

Guo, Y., Kim, S.S., Timothy, D.J. and Wang, K.C. (2006) Tourism and reconciliation between Mainland China and Taiwan. *Tourism Management* 27 (5), 997–1005.

Griffiths, M. (2016) Hope in Hebron: The political affects of activism in a strangled city. *Antipode* 49 (3), 1–19.

Grimwood, B.S.R., Stinson, M.J. and King, L.J. (2019) A decolonizing settler story. *Annals of Tourism Research* 79, 102763. https://doi.org/10.1016/j.annals.2019.102763.

Habermas, J. (1984) *Theorie des kommunikativen Handelns*. Frankfurt: Suhrkamp Verlag.

Higgins-Desbiolles, F. (2006a) More than an 'industry': The forgotten power of tourism as a social force. *Tourism Management* 27 (6), 1192–1208.

Higgins-Desbiolles, F. (2006b) Reconciliation tourism: Healing divided societies. IIPT Occasional Paper No. 7. See www.iipt.org/educators/OccPap07.pdf (accessed May 2016).

Higgins-Desbiolles, F. (2008) Justice tourism and alternative globalisation. In Palestinian Alternative Tourism Group (ed.) *Combating Dispossession: Towards a Code of Ethics for Tourism in Palestine* (pp. 345–364). Beit Sahour: ATG.

Hobson, J. and Ko, G. (1994) Tourism and politics: The implications of the change in sovereignty on the future development of Hong Kong's tourism industry. *Journal of Travel Research* 32 (4), 2–8.

Hollinshead, K. (2009) The "Worldmaking" prodigy of tourism: The reach and power of tourism in the dynamics of change and transformation *Tourism Analysis* 14 (1), 139–152. https://doi.org/10.3727/108354209788970162.

Inui, Y., Wheeler, D. and Lankford, S. (2006) Rethinking tourism education: What should schools teach? *Journal of Hospitality, Leisure, Sport and Tourism Education* 5 (2), 26–36.

Isaac, R.K. (2010a) Moving from pilgrimage to responsible tourism: The case of Palestine. *Current Issues in Tourism* 13 (6), 579–590.

Isaac, R.K. (2010b) Palestinian tourism in transition: Hope, aspiration, or reality. *The Journal of Tourism and Peace Research* 1 (1), 16–26.

Isaac, R.K. (2018) Religious tourism in Palestine: Challenges and opportunities. In R. Butler and W. Suntikul (eds) *Tourism and Religion: Issues and Implications* (pp. 143–160). Bristol: Channel View Publications.

Isaac, R.K. (2021) Pilgrimage tourism to Palestine: The Come and See initiative in Palestine. In D. Liutikas (ed.) *Pilgrims: Values and Identities* (pp. 188–201). Wallingford: CABI.

Isaac, R.K. and Ashworth, G. (2012) Moving from pilgrimage to dark tourism: Leveraging tourism in Palestine. *Tourism, Culture and Communication* 11 (3), 149–164.

Isaac, R.K. and Hodge, D. (2011) An exploratory study: Justice tourism in controversial areas. The case of Palestine. *Tourism Planning & Development* 8 (1), 101–108.

Isaac, R.K. and Platenkamp, V. (2012) Ethnography of hope in extreme places: Arendt's agora in controversial tourism destinations. *Tourism, Culture & Communication* 12, 173–186.

Isaac, R.K. and Platenkamp, V. (2016) Concrete U(dys)topia in Bethlehem: A city of two tales. *Journal of Tourism and Cultural Change* 14 (2), 150–166.

Isaac, R.K., Hall, C.M. and Higgins-Desbiolles, F. (2016) *The Politics and Power of Tourism in Palestine*. Abingdon: Routledge.

Kassis, R.O. (2011) *Kairos for Palestine*. Ramallah: Badayl/Alternatives.

Kassis, R., Solomon, R. and Higgins-Desbiolles, F. (2016) Solidarity tourism in Palestine: The alternative tourism group of Palestine as a catalyst instrument of resistance. In R. Isaac, C.M. Hall and F. Higgins-Desbiolles (eds) *The Politics and Power of Tourism in Palestine* (pp. 35–52). Abingdon: Routledge.

Kim, Y. and Crompton, J. (1990) Role of tourism in unifying the two Koreas. *Annals of Tourism Research* 17 (3), 353–366.

Lisle, D. (2000) Consuming danger: Reimagining the war/tourism divide. *Alternatives: Global, Local, Political* 25 (1), 91–116.

Litvin, S. (1998) Tourism: The world's peace industry? *Journal of Travel Research* 37 (1), 63–66.

Lueck, M.A. (2007) Hope for a cause as cause for hope: The need for hope in environmental sociology. *American Sociologist* 38 (3), 250–261.

Mannheim, K. (1936) *Ideology and Utopia*. Madrid: Mariner Books.

McGeer, V. (2004) The art of good hope. *The Annals of the American Academy of Political and Social Science* 592, 100–127.

Mitchell, L. and Murphy, P. (1991) Geography and tourism. *Annals of Tourism Research* 18 (1), 57–70.

Moynagh, M. (2008) *Political Tourism and Its Texts*. Toronto: University of Toronto Press.

Noy, C. (2011) The political ends of tourism: Voices and narratives of Silway/the city of David in East Jerusalem. In I. Ateljevic, N. Morgan and A. Pritchard (eds) *The*

Critical Turn in Tourism Studies: Creating an Academy of Hope (pp. 27–41). Abingdon: Routledge.

Ojala, M. (2012) Hope and climate change: The importance of hope for environmental engagement among young people. *Environment Education Research* 15 (5), 625–642.

Patton, M. (2002) *Qualitative Research and Evaluation Methods*. Thousand Oaks: SAGE.

Pearce, D.G. (1992) Alternative tourism: Concepts, classifications and questions. In W.R. Eadington and V.L. Smith (eds) *Tourism Alternatives: Potentials and Problems in the Development of Tourism*. Philadelphia: University of Pennsylvania Press.

Pitts, W.J. (1996) Uprising in Chiapas, Mexico: Zapata lives – tourism falters. In A. Pizam and Y. Mansfeld (eds) *Tourism, Crime and International Security Issues* (pp. 215–227) Chichester: Wiley.

Richard, A. and Rudnyckyj, D. (2009) Economic of affect. *Journal of the Royal Anthropological Institute* 15 (1), 57–77.

Richter, L. (1994) Political dimensions of tourism. In J. Ritchie and C. Goeldner (eds) *Travel, Tourism and Hospitality Research: A Handbook for Managers and Researchers* (pp. 219–231). New York: Wiley.

RT (2014) European Parliament votes to recognize Palestine statehood 'in principle'. *RT*, December 28. See www.rt.com/news/215227-eu-recognize-palestine-state. (Accessed June 2021)

Sarkar, S.K. and George, B.P. (2010) Peace through alternative tourism: Case studies from Bengal, India. *The Journal of Tourism and Peace Research* 1 (1), 27–41.

Selwyn, T. and Karkut, J. (2007) The politics of institution building and European cooperation: Reflections on an EC TEMPUS project on tourism and culture in Bosnia-Herzegovina. In P. Burns and M. Novelli (eds) *Tourism and Politics: Global Frameworks and Local Realities* (pp. 123–145). London: Elsevier.

Scheyvens, R. (2002) *Tourism for Development: Empowering Communities*. Harlow: Prentice Hall.

Shepherd, J. and Laven, D. (2019) Providing counter-narratives: The positive role of hostels in the Israeli-Palestinian context. *Tourism Geographies* 22 (4–5), 848–871. https://doi.org/10.1080/14616688.2019.1669215.

Tabar, L. and Salamanca, J. (2015) After Oslo: Settler colonialism, neoliberal development and liberation. In *Critical Readings of Development under Colonialism: Towards a Political Economy for Liberation in the Occupied Palestinian Territories* (pp. 9–32).Centre for Development Studies. Birzeit, Palestine: Birzeit University and the Rosal Luxemburg Stiftung.

Timothy, D., Prideaux, B. and Kim, S.S. (2004) Tourism at borders of conflict and (de) militarised zones. In T.V. Singh (ed.) *New Horizons in Tourism: Strange Experiences and Stranger Practices* (pp. 83–94). Wallingford: CABI.

Tucker, H. (2016) Empathy and tourism: Limits and possibilities. *Annals of Tourism Research* 57, 31–143.

Tucker, H. and Shelton, E. (2018) Tourism, mood and affect: Narratives of hope and loss. *Annals of Tourism Research* 70, 60–75.

Twain, M. (2007) *The Innocents Abroad, or The New Pilgrim's Progress*. New York: Signet Classics.

Urry, J. (1990) The consumption of tourism. *Sociology* 24 (1), 23–35.

Var, T., Schluter, R., Ankomah, P. and Lee, T. (1989) Tourism and peace: The case of Argentina. *Annals of Tourism Research* 16 (3), 431–434.

Van Amerom, M. and Buscher, B. (2005) Peace parks in Southern Africa: Bringers of an Africa Renaissance? *Journal of Modern Africa* 43 (2), 159–183.

Van den Boer, D. (2016) Toward decolonization in tourism: Engaged tourism and the Jerusalem tourism cluster. *Jerusalem Quarterly* 65, 9–21.

Warner, J. (1999) North Cyprus: Tourism and the challenges of non-recognition. *Journal of Sustainable Tourism* 7 (2), 128–145.
Willis, E. (2014) *Theatrically, Dark Tourism and Ethical Spectatorship*. New York: Palgrave Macmillan.
White, B. (2015) 'Visit Palestine' says West Bank's growing alternative tourism industry. *Electronic Intifada*, 20 July. See https://electronicintifada.net/content/visit-palestine-says-west-banks-growing-alternative-tourism-industry/8343 (accessed May 2016).
Yu, L. (1997) Travel between politically divided China and Taiwan. *Asian Pacific Journal of Travel Research* 24 (1), 42–44.
Zournazi, M. (2002) *Hope: New Philosophies for Change*. London: Psychology Press.

13 Deep Peace and the Solo Wilderness Canoe Experience

David A. Fennell

Introduction

The canoe played a pivotal role in shaping the cultural and ecological landscapes of North America. Long before Europeans arrived in the New World, the canoe provided the means for Turtle Islanders (the name Algonquian- and Iroquoian-speaking peoples of the north-east gave to North America) to penetrate areas that they would not otherwise access in a land defined by a vast network of rivers and lakes. Likewise, the canoe was essential for white settlers to move goods from remote regions to European markets, build and connect communities, and spread religion throughout the New World. The French Jesuit Missionary Father Jean de Brébeuf, who arrived in the New World in 1625, spent most of his life living with the Huron Indians, learning their language and documenting their lives. We learn more about the rigours that Brébeuf endured through his journals. In one such account, Brébeuf wrote that he was 'sometimes so weary that the body could do no more, but at the same time [his] soul experienced very deep peace, considering that [he] was suffering for God; no one knows it if he has not experienced it' (Kenton, 1954, cited in Franks, 1977: 206). In the spirit of Brébeuf, Franks (1977: 205) writes that modern recreational canoeists might also 'experience a deep peace in the midst of fatigue. It comes to him for the same reason that it came to Father Brébeuf'.

From Brébeuf's and Franks' accounts, a profound sense of peace comes from serving spiritual and recreational interests. However, given the history of deep peace tied to canoeing, what do we know about it, and how does it relate to canoeing or other forms of tourism and outdoor recreation (TOR) in the present? A review of the literature shows that a formal definition of *deep peace* presently eludes us; however, *peace* has been defined as a 'transitory and dynamic state of mind related to acceptance, contentment, and surrender' (Røgild-Müller & Robinson, 2020), or a harmonious state of happiness (Xu *et al.*, 2014). Furthermore, scholars argue that peace as a state of mind is hard to pin down because it shows variability across time and cultures and between individuals, indicating a deeply subjective nature (Valsiner, 2001).

Peace is emerging as a popular (and important) topic of research in tourism but studies have adopted an almost exclusive focus on a broad or macro-level approach, as in peace as a vehicle for intercultural understanding, justice, and human rights as a matter of security and economic development (Guasca *et al.*, 2022) Examples of this macro-level application of peace include Moufakkir and Kelly's (2010) edited book, which includes several case studies on peace; Pratt and Liu's (2016) global view questioning whether or not tourism is a conduit of peace; and the work of Ym (2022), who analysed the relationship between tourism, violence, and conflict. A recent (2022) special edition of the *Journal of Sustainable Tourism* includes a number of articles that embrace this macro-level perspective on how peace studies can advance the overturning of structural injustices, reconciliation and decolonisation in addressing wider crises and challenges (Higgins-Desbiolles *et al.*, 2022).

In sharp contrast, this chapter investigates peace from a micro-level or atomistic perspective pertaining to peace as a deeply subjective state of mind or state of happiness, as noted in the definitions just provided. A comprehensive understanding of deep peace may open the door to a more detailed awareness of existential moments in TOR activities like wilderness canoe experiences that link back to the historical reference points of de Brébeuf and Franks. Deep peace may therefore provide new conceptual, theoretical, and empirical insights for such moments and the broader events in which these moments are situated. As such, the purpose of this chapter is to identify the fundamental characteristics of deep peace through a seven-day solo canoe trip to the Temagami region of northern Ontario, Canada, in July of 2021. Personal narrative experience is the autoethnographic approach adopted in achieving this end.

Literature Review

There are two main sections to the review of the literature. The first explores the canoe literature, both academic and in popular culture. The second reviews how the terms 'deep' and 'peace' have been used in select TOR journals, along with subjective well-being and mindfulness as related concepts.

The nature of canoeing

Several magazine articles and popular culture books provide detailed accounts of solo (and group) canoe trips that discuss canoeist motivations and emotions. For example, a survey on the solo canoe experiences of

Facebook readers of *Family Camping & Canoeroots* (Anon, 2016) magazine yielded comments such as 'frightening', 'inspiring', 'empowering', 'soothing', 'spiritually enriching' and 'impossibly lonely'. Hoyle (1999) writes that there is fear of death, starvation, being caught in cold weather, being lost, accidents, animal attacks, and running out of food (Hoyle, 1999; see also Raffan, 1999b). Ross's (1995) solo canoe trip across Canada was punctuated by periods of profound peace. On the first leg of his trip, Ross wrote, 'I like the sense of isolation I'm starting to feel' (1995: 22). For Franks (1977: 175), solo canoeing has its pleasures in the form of 'nature-watching, fishing, solitude, introspection and extrospection'.

Connections have also been made between the canoe trip and Daoist meditation – that is, streamings of the energy of the universe based on harmony with nature (Horwood, 1999). Emptiness, effortless action and the union of polarities referenced in Tai Ji correspond to the four aspects of canoeing: 'the form of the canoe, the act of paddling itself, the nature of rivers and, that quintessential union of them all, the canoe trip' (Horwood, 1999: 65). The temporal aspect of Csikszentmihalyi's (1990) concept of flow enters Horwood's (1999) discussion of Daoism, where canoeists perform the mechanics of paddling self-consciously. The aspect of 'meditation in motion', which surfaces in work by Thomas (1999: 176), provides a sense of self-renewal and a 'deep sense of being – a spiritual connection, an immersion'. For Thomas, the canoe trip is as much an inner journey as it is an adventure. Research has also highlighted the canoe trip as a religious quest or pilgrimage representing the transitory nature of man on earth: the canoeist's journey, like life itself, has a beginning (departure), middle (trials and tribulations of the trip) and end (James, 1981).

The multidimensional nature of the canoe trip is confirmed by Benidickson (1997), who identifies several characteristics, including love of landscape, being closer to God, connecting to ancestors, finding and healing ourselves, physical and emotional renewal and unrestricted mobility. For Raffan (1999a: 34), the canoe trip nurtures awareness and 'situated knowing'. Still others argue that the canoe trip summons the latent biological being that allows us to become wilder (Henderson, 1999), tapping into our inherent love of and connection to nature (biophilia), which prompts us to want to conserve and preserve wild places (Grimwood, 2011; see also Mullins, 2009). Canoe research has also focused on recreational carrying capacity, including, for example, encounter norms (Lewis *et al.*, 1996), as well as female-run, single-gender canoe programmes, which increase women's sense of physical confidence, strength and competence in challenging their own identities (McDermott, 2004).

Scholars have also investigated the canoe as a conduit for ecological and sociocultural destruction on campsites, through the consumption of fuel to get canoeists to their destinations (Hodgins, 1988). Grimwood (2015; see also Erickson, 2008) identified the many problems around the colonial domination of the Canadian Arctic through canoe tourist interpretations of nature and culture. Leisure is part of this capitalist development where this 'freedom to be' (citing Kelly, 1987: 238) is 'not an escape from the anxieties of capitalism, but a collaboration with it' (Erickson, 2008: 183). Referring to the canoe experience as the domain of white middle-class men offers the opportunity to perform masculine identities (see Fletcher, 2014; Haun-Moss, 2002), while people of colour, including Indigenous people, were welcome in the canoe culture of the late 1800s and early 1900s not as participants but rather as laborers (Dunkin, 2019). In the same vein as Grimwood (2015), recent work has stretched the boundaries of scholarly discourse on the canoe through the argument that canoeing is just as much a factor in colonialism, activism, resource politics, nationalism and governance as it is a form of wilderness recreation and nation building (Erickson & Wylie Krotz, 2021).

Searching for deep peace

A comprehensive search of Google Scholar and the web more generally yielded few references to the concept of deep peace. Deep peace has been captured in Danish song (Andersen, 2006), as a Gaelic blessing (Rutter, 1978) and in the introduction (written by William Frederic Badè) of John Muir's book, *A Thousand-Mile Walk to the Gulf* (Badè, 1916: x). Phipps (1987: 32) articulated a philosophy of deep peace as 'fortitude of soul' and 'strength of heart' that can manifest itself in a type of metaphysical liberation in which nothing can prevail against it. Furthermore, he argues that thinking rationally – that is, thinking dualistically: light/dark, east/west, good/evil – elevates cerebral cleverness over wisdom, the latter of which is grounded in 'feelings, sensations and intuitions' and is far closer to awareness. Inner wisdom (e.g. awareness of oneself) can manifest as a form of deep peace.

Apart from these few mentions, deep peace is virtually non-existent in the academic literature. Therefore, to build a greater understanding of the principal components of deep peace (i.e. 'deep' and 'peace') I endeavoured to understand better how these terms are being used by reviewing the full content of journals related to the focus of the present study: *Journal of Ecotourism, Journal of Outdoor Recreation and Tourism, The Journal of Environmental Education* and *Journal of Experiential Education*. As space prevents a full description of these studies, Table 13.1 summarises this information according to theme, author, and journal.

Concepts such as mindfulness and subjective well-being have been used to understand the psychological nature of TOR experiences. Mindfulness relates to the regulation of attention to immediate experiences 'characterised by curiosity, openness, and acceptance' (Bishop *et al.*, 2004: 232), leading to intense feelings of satisfaction (Moscardo, 2009). Mindfulness can be achieved in slow adventure pursuits through the choreographing of time, meaningful moments, and social cohesion by guides leading to

Table 13.1 Peace and Deep themes in the literature

Themes	Author	Journal
Peace		
Peace, democracy and justice	Maina-Okori et al. (2018)	JEE
Peace-building or Peace Corps	Mueller Worster (2017)	JEE
Peace parks	Chirozva (2015)	JOE
Peace and tranquillity or peace and quiet	Lindholst et al. (2015)	JORT
Peace at ecolodges	Lai & Shafer (2005)	JOE
Great white pine as a symbol of peace	Caduto (1998)	JEE
Wild nature as peace	Beckford et al. (2010)	JEE
Peace through journaling in nature	Tsevreni (2021)	JEE
Deep ecology as peace	Kowalewski (2002)	JEE
Peace as psychological benefit of being in nature and part of a larger system	D'Amato & Krasny (2011)	JEE
Peace from being a responsible tourist	Weeden (2011)	JOE
Peace as existential and psychological well-being through hiking	Hitchner et al. (2019)	JOE
Recreation experience preference scales: peace and tranquillity	Lepp & Herpy (2015); Vistad et al. (2020)	JORT, JORT
Deep		
Nature providing deep reflection	Ingólfsdóttir & Gunnarsdóttir (2020)	
Nature providing a deep sense of place as geopiety	Knowles (1992)	JExE
Deeper relationships with others on long sailing trips	Jirásek & Hurych (2019)	JORT
Deep views on wilderness equity through hunting	Eliason (2016)	JORT
Deep enjoyment in seeing wild animals	Curtin (2005)	JOE
A deep understanding of the natural world makes it a priority in our lives	Chakraborty (2019)	JOE
Strong Deep/Peace Connections		
A deep sense of peace through trail riding (on horseback) sets the mind at rest	Sawchuk (2016)	JOE

JEE = The Journal of Environmental Education; JExE = Journal of Experiential Education; JOE = Journal of Ecotourism; JORT = Journal of Outdoor Recreation and Tourism.

hedonic and eudaimonic outcomes in deepening the experience for tourists (see Farkić et al., 2020). Additionally, subjective well-being (SWB) is a measure of happiness and life satisfaction that moves away from economic determinants and towards longer-term feelings of joy and happiness that play a part in one's positive (or negative) psychological health (Diener et al., 1999; McCabe & Johnson, 2013). Scholars have found that SWB is enhanced through nature-based tourism activities that allow for 'enjoying

[the] natural environment and escaping from daily life', 'pursuing intimacy' and 'pursuing a healthy life' (Kim et al., 2015: 585).

Other studies have used the concept of capital (Bourdieu, 1977) and its different forms to better understand TOR participation. Capital is intricately tied to power, ranking and advantage in society within the spaces in which it resides. This includes cultural capital or the social assets of a person (Newbery, 2003), social capital or 'who you know' (Adler & Kwon, 2002), human capital or 'what you know' (O'Leary et al., 2002) and psychological capital or 'who you are' (Luthans et al., 2007). This latter form is characterised by self-confidence to succeed at tasks, perseverance to achieve goals and resilience in the face of challenges to succeed (Luthans et al., 2007). Though this was done outside the boundaries of canoe tripping, Nettleton (2013) introduced a new form of capital – existential capital – to explain that pleasure may be experienced in physically exhausting activities like fell running. Nettleton argues that, for the fell runner, the existential embodied (mind/body) experience generates a resource in the form of well-being rooted in the link between the corporeal self and the physical landscape. An 'experience is so deeply personal, it can only be shared and appreciated by others who have done it' (Nettleton, 2013: 206).

Study Region and Methodology

At over 2000 km², Temagami, Ontario, Canada, approximately five hours north of Toronto, is home to the world's largest stand of old-growth white and red pines and was the centre of environmental battles at various points in recent history to save these forests from logging. The traditional people of the region, the Te-mee-ah-gamaw anishinawbeg or 'Deep Water People', have for 5000 years travelled one of the world's largest winter and summer trail networks (1300 trails in total) (Wilson, 2011). These trail systems and the geology of the Canadian Shield, countless carved out river and lake systems in a transition zone between boreal forest (north) and the Great Lakes-St. Lawrence forests (south), make Temagami a premier canoe destination (Wilson, 2011).

The methodological approach adopted in this study is autoethnography. While many scholars view autoethnography as a methodology (Wigglesworth, 2018), others see it more as an epistemological orientation because 'the relationships among experience, knowledge, and representation' have several methodological implications (Butz, 2010: 139). Autoethnography challenges ethnographic authority's ontological and epistemological foundations by recognising that knowledge generation can be subjective and can straddle emotions and experience (Butz & Besio, 2009). Defined, autoethnography is 'a form of self-narrative that places the self within a social context' (Reed-Danahay, 1997: 9), or, as explained by Butz and Besio (2009: 1662), the representational outcome – the performance, in a sense – of a process of *critical reflexivity*, where reflexivity underscores the need to examine how our own beliefs and practices influence the research process (Finlay, 1998).

Butz and Besio (2009) identify five different autoethnographic practices that exist along a continuum. These include personal experience narrative, narrative ethnography, insider or complete member research, Indigenous autoethnography, and subaltern autoethnography. The practice that conforms most closely to the present study is personal narrative experience, which Butz (2010: 139) defines as 'academics who study their own life experiences intensely in order to illuminate a larger social or cultural phenomenon'. Furthermore, autoethnographic data can be analytic or evocative. In the former case, data is subject to insight and analysis that, although rigorous, may ultimately transform and lose the meaning of the data (Ellis & Bochner, 2000). On the other hand, evocative data is characterised by 'deep self-analysis' (Kennedy *et al.*, 2019: 190) based on self-reflexive, subjective, personalised lived experience narratives designed to arouse emotional and empathetic responses in readers (Ellis & Bochner, 2000).

Data collection took place through journaling, which allows for 'richly textured and deeply felt renderings of experiences' (Wigglesworth, 2018: 792) – an approach that corresponds well to seeking deep peace. Furthermore, a solo wilderness canoe trip is cultural practice because it aligns with middle-class values and 'whiteness' (Fletcher, 2014). I engage this project as one who fits within this categorisation, which influences how I see the world, and my interpretation and understanding of the literature.

Narrative thematic analysis (Cooper & Lilyea, 2022) was used to identify broad themes from my journal. Cooper and Lilyea (2022), Gibbs (2008) and Janesick (2010) argue that there are many thematic analysis tools that may be used in autoethnographic research. I chose the development of a coding scheme that follows Braun and Clarke (2006), which included (1) becoming familiar with the data (journal entries), (2) generating initial codes, (3) generating initial themes, (4) reviewing themes against coded data and the dataset in general, (5) defining and naming themes and (6) providing a thick description of the results. Braun and Clarke (2006) argue that thematic analysis is noted for its flexibility and not tied to a particular paradigmatic orientation – post-positivist, constructivist or critical realist.

Results

Table 13.2 illustrates the breadth of primary words and phrases that were extracted from the many journal entries recorded throughout the trip, culminating in eight broad themes (see also Figure 13.1). The first of these, Connections, is based on 19 different words and phrases and four subthemes. The first of these, connections to self, included a voyage into the inner self, the personal quest of the hero and aspects of identity. Connecting to family emerged as important for me by thinking about deep relationships

Deep Peace and the Solo Wilderness Canoe Experience 209

Table 13.2 Deep peace themes from the solo canoe trip

Primary words/phrases: Open coding	Sub-themes: Axial coding	Theme: Selective coding
Voyage into inner self Not needing to be the "hero" Challenge my identity as a person/canoeist Reaffirmation of identity Extension of identity	Connecting to self	1. Connections
Disconnecting with others Deep relationship with others Primaeval connection to wilderness Past connections with Indigenous people Historical connections (annual trips, retracing routes with family, explorers)	Connecting to other people & history	
Passive thoughts to spirituality The trip as a pilgrimage Canoeing as meditation in motion Spiritual health	Connecting to God/spirituality	
Love of nature; Romanticising about nature Nature as 'real' Harmony and bonding with nature Deep views on wilderness equity Reconnections with nature after time away	Connecting to nature	
Death Cold Being lost Accidents Animals No food Absence of fear Importance of being able to cope with stress Conditions of the water and weather	Emotions tied to fear and stress	2. Affective domain
Awe Happiness	Emotions tied to joy	
The experiencing peace and freedom Experiencing peace and humility Magic in peace and solitude Quiet, linked to peace	The experience of peace	
Feelings of loneliness Liking isolation Heightened sense of awareness Feeling empowered Feeling inspired Feeling soothed Feeling better about myself Feeling or experiencing pleasure/pain Feeling privileged to be here	Negative and positive feelings or states	
Peace from physical and emotional challenges Satisfaction of hardships endured (e.g., lifting packs and portaging) Gaining confidence and strength from physical tasks	Physical achievement	3. Achievement and Skill

(Continued)

Table 13.2 (*Continued*)

Primary words/phrases: Open coding	Sub-themes: Axial coding	Theme: Selective coding
Sense of accomplishment Triumph of personal challenges Matching abilities with challenges Intrinsic rewards from participation Extrinsic reward for successful trip.	Mental/emotional achievement	
Satisfaction in proper movement and skill Being able to read the land Magic in the feel of paddle and canoe Not being overconfident about one's skills	Development of skills	
Learning of lessons Deep introspection of educational practices Knowing how to read the signs & environment Changing conditions of the environment Gaining more intimate knowledge of Temagami	Learning and education	4. Learning and Change
Reorientation of values Inner change Changing over time	Examination of one's values	
The use of only simple material affordances (post-materialism) Embracing material goods in enhancing the experience (bodily fix) The need to buy the best equipment The pleasure of buying goods for the canoe trip Shameful conspicuousness connected with new apparel Dilemmas of using old versus new equipment	Materialism and post-materialism	5. Materialism, Equipment & Planning
Fussing over gear Experimentation with new clothes and gadgets Importance of trip planning Taking and using maps Taking the right type and amount of food Taking the correct type of equipment Several days needed to plan the trip Packing several times to cut weight Being thoughtful about gear in the absence of partner Using gear purchased over time: utilitarian argument	Pre-trip planning and equipment	
Awareness and situated knowing The need to be flexible during trip Being vigilant with map and compass	On-trip equipment & planning	
My involvement in Temagami as a white, middle-class male in reference to capitalism and colonialism Disturbance of resources & indigenous cultures Pride in Temagami; ours to share, enjoy and be responsible for together	Historical legacies	6. Responsibility
Letting go of urban encumbrances Making sure others are safe at home Reconciling obligations/responsibilities at home and work (missing loved ones) Passing traditions and knowledge on to family Pilgrimage to honour family tradition and socialisation	Responsibilities of home/family	

(*Continued*)

Table 13.2 (*Continued*)

Primary words/phrases: Open coding	Sub-themes: Axial coding	Theme: Selective coding
The peace I found in trying to be a responsible tourist Being sustainable Ethical purchasing Working to do my part to keep as pristine as possible through my responsible behaviour	Responsibility in action	
Space The importance of getting away Unrestricted mobility afforded by the canoe Escape from civilisation (schedules, watches) Peace as part of a larger system	Spatial aspects	7. Space, Place and Time
Place Return to Temagami as a perceived paradise The need to have a first-hand engagement with Temagami The need to have a deeper exploration of Temagami The idea of 'North'	Place and reverence for place	
Time Bing mindful of time too often Wishing I had more time Making up time because of conditions Mindfulness that could be achieved through slow adventure pursuit	Temporal aspects	
The canoe trip to Temagami as being healthy for me psychologically Therapeutic; physical and mental renewal The daily need to eat well. The mindset that the canoe trip is a form of fitness	Maintaining a healthy diet and practices	8. Health & Safety
Changing techniques to stay safe Where to camp Doing things slowly and deliberately The use of new technologies to be safe	Remaining safe in a wilderness setting	

and historical connections with past family trips. One of the first thoughts that entered my mind after the passing of two loved ones in 2017 was to embark on a solo canoe trip to put everything into perspective. Although life got in the way in the ensuing years, the time finally came in 2021 when I was ready to follow through with what I wrote in February 2017 to put the pieces back together again:

> It's time to hitch up the wagon and go but know you can always come home.
>
> Because going out, it has been said, is really going in. . . .
>
> And we do need to move forward, unwillingly at first.
>
> With the knowledge that they would want that.
>
> An unspoken promise at the end that speaks of continuity.

I found value in that Connections also meant reconnections with a dormant former self. The trip allowed me to explore the natural world in a way that I had not done in years. Slowly I reacquainted myself with the biodiversity that was all around me. I became curious to identify and understand but also frustrated over my inability to recall what I had learned many years ago as a student. Such I attribute to the opportunity and willingness to try and slow down and 'be' with nature without the encumbrances of my world.

> Love of nature through the lens that was so important to me many years ago was replaced by the love of family and all that goes along with that. I was pleasantly surprised on this trip at how motivated I was to photograph trees, insects, fungus, and other forest treasures.

Being alone in the wilderness provided the opportunity to experience many different psychological states, from basic emotions like fear and joy to feelings of peace, isolation, and states broadly classified as positive or negative. Although I have plenty of experience on canoe trips, the one fear I consistently maintain is of black bears. I would not call this a deep fear, but it is there nevertheless. From my notes:

> On the second day, I encountered a black bear feeding on berries close to the water's edge as I approached a portage in a quiet bay. I managed to get about 20 m from the bear and take four photographs before it saw me and bolted up the hill. It showed me how scared black bears could be of humans and quelled any fears of bears from there on in.

The sense of awe and happiness, categorised as joy, was evident in the trip, especially given the anticipation built for the experience over several months. Such was not a sense of awe that accompanies the experience of seeing things for the first time but rather the renewal of awe in witnessing again the world's largest stand of old-growth white and red pines. Temagami is not alone as a case study in the fight to save wild places, but it is a story worth remembering and telling to a country and world that continue to slip into a technological, commercial, and urbanisation coma that stifles the human–wilderness bond.

Not unlike the experience of other solo canoeists identified in the review of literature, I experienced feelings of peace (and freedom) and isolation. I liked that I did not have to answer to other people in organising the day, meals, where to place the tent, and other details that would have been shared. However, I also enjoyed the feeling of being alone and how this peace and quiet was so different from life at home:

> One night, I woke at 12:38am and forced myself to stay up for at least 10 minutes to listen. There was no noise at all, not even wind. Then, after about 10 minutes, I heard a faint call/cry many km away. It sounded first like the high-pitched whistle of a locomotive. The third time I heard it, it was more audible, and I recognised it as a wolf.

I also felt the oscillations between pleasure and pain. The feelings of pleasure far outweighed the pain despite the intensity at specific points.

The canoe trip is equal parts pleasure and pain. The pain and weight of the canoe on your traps with a pack on your back. The pain of canoeing for sometimes seven hours of the day and the cramping while sitting and kneeling in the boat. Almost a blessing to get to a portage to stretch the legs.

Connected to the pain and pleasure of the canoe trip was a sense of Achievement and Skill (another theme). Examples included the pleasure of actually completing the trip, flow in terms of matching skills with challenges, and intrinsic rewards from participation. However, the idea of an extrinsic reward surfaced as I got closer to the end of the trip as a symbol of remembrance:

Given the trip's difficulty, I feel the need to reward myself with a small trip token. Maybe a bracelet as a marker of remembrance. A marker of success. However, how do you measure success, and why measure it at all?

The trip was also an opportunity to hone skills developed over time and test these skills under conditions that were not always favourable. I was satisfied with my skill level in moving the canoe in needed directions, but I also felt that more trips would enable me to increase my skills and avoid being overconfident or failing to appropriately link skills with challenges. Such was evident in my first strokes of the trip, where eagerness took the place of reason.

Rookie move to not stick closer to shore on a bay that was windy with whitecaps. I was just focused on getting started with the trip after many weeks of planning and a long day of getting to the water. It was probably the second-worst day of canoeing I can remember. The headwind was strong, and the whitecaps (one and two-foot) were sometimes so high as to spill over the canoe's bow.

In reference to the theme of Learning and Change, as I moved through the week, I felt the need for a more intimate knowledge of Temagami. The education and learning transmission wheels started to turn for me as I recognised that much more could be done to educate canoeists about the natural and cultural history of Temagami, mainly based on my observation of YouTube and Vimeo video clips of other canoeists in my preparation for this trip. I wrote:

There is scope to understand better the flora, fauna, geology, weather, hydrology, people of the region. I found myself captivated by the idea of working on a video series the following summer . . . to slow down and more thoroughly understand the natural history of the region deeply.

Change was not just about the conditions of Temagami but also an inner change of values over time. How I approach a canoe trip now is very different from how I approached these adventures 20 and 40 years ago. There are watershed moments in life – markers – that define us mentally, emotionally, and physically. A solo trip to Temagami, for me, was one such moment.

> Maybe the marker should be change as I move through the years and get closer to 60. Being more thoughtful, aware and less reckless should be the goal. Using a lighter canoe and packs. Portaging a trail between lakes twice, instead of always once to spread out the load, using lighter gear and having the right gear is more important as I age and not having to be the Hero.

I now imagine that leaving the heroic past favouring a measured future is my only option as a solo canoeist moving forward.

Although initially thought to be too broad to include in the same theme, there were many similarities between codes and sub-themes in the Materialism, Equipment and Planning theme. The sub-theme of materialism and post-materialism provided insight into my preparation for this trip and my behaviour as an outdoor recreationist. I have leaned towards a hybrid model between the need/want to purchase the best equipment – indeed, I gain great pleasure in the purchase of these new products, but

> [e]ven as I purchase better, lighter gear, I return to my old tripping clothes that I've used for almost 30 years. I carry with me new high-tech clothes that are light, but these clothes typically stay in the pack for emergencies.

Pre-trip planning, the second sub-theme, represented the most significant investment in time, as illustrated in the quote above on buying good equipment. It links with the section on Connections, where the trip plan was starting to take shape after my father's death. But as the trip drew near, preparation activities intensified through fussing over gear, experimentation with new clothes and gadgets, assembling maps and gear – the need to get it right by:

> [p]acking my gear three times, eliminating gear/food each time. I had wanted to take one big pack but knew that taking two smaller packs, although more challenging to portage, would be better to trim out the canoe (balance the canoe from bow to stern). Weight is the enemy!

The value of planning before the trip was mirrored by the importance of planning while on trip, even though this was not the case in the literature. The other aspect that surfaced was being vigilant with map and compass. Being alone means there is no one else to bounce ideas and thoughts off when reading maps, so choosing routes and distances carefully is essential for the solo paddler.

Of the three sub-themes of Responsibility, home and family responsibilities and responsibility in action were more important than historical legacies. While the latter category is of considerable importance, I felt a sense of pride that this place is ours to share, enjoy, and be responsible for together. Still, it made me think of a student trip to Temagami many years before when we found our way onto an Indigenous man's property, where

> [w]e offered a small gift to him as a sign of respect, and I felt that this gesture would stay with the students forever as a responsible manner in which to interact with First Nations people, especially as temporary occupants of land that is so intricately tied to the indigenous way of life.

What occupied my thoughts more in this theme was responsibilities for home and family, especially about obligations at home and the selfishness of being away.

> Part of me felt selfish about being away from family. However, I realised that my family is not interested in canoe tripping anymore and that if I wanted to canoe, I would have to do it by myself. I've always put family first. I guess it's time to start returning to a focus on me as kids leave the nest.

There was also recognition that being responsible means being active in treating the natural world with respect. Although this is a topic of research and teaching for me, meaning it is always front of mind, the trip provided an opportunity to trace back to formative years:

> On three campsites, there were old rusty cans beside the firepit. It reminded me of the canoe trips taken with family and my dad insisting on bringing a big can of Irish stew. After cooking the stew, he would burn (the label), bash (the can), and then bury it, which was common campsite ethics for the day. In 2021, leaving cans or anything else behind is severely frowned upon, so I packed some trash out.

I found considerable peace in responsibility in my efforts to leave this well-reputed destination in good shape for future occupants: leaving firewood for the next occupant, making sure no food waste is left on shores, and picking up after others who used campsites before me.

The ruggedness and northern location of Temagami suggests a friction of distance factor that plays heavily in the choice of where to canoe. This factor relates to the idea of 'North' that is important to many Canadians. However, it was not just the quantitative aspects of space but, probably more critically, the qualitative notions of 'place' that were essential in defining the essence of my trip:

> I have often felt that the idea of North is a symbol of who we are as a people. The setting, activity and being alone was like a catalyst for thinking about the value of this unique region, crystallised through the frequent call of loons and fresh air.

While space and place emerged as essential considerations on the trip, time was even more critical as I struggled with older conceptions of the canoe trip experience. For me, there is a constant need to move through the trip quickly. By the end of this trip, however, I felt the need to slow down to get to know Temagami more intimately in future trips because there is the sense that something is left behind or that I have been cheated out of the more personal and unique attributes of Temagami that I have been unable to find – not seeing the forest for the trees.

This preoccupation with time was especially evident on days where the weather (wind and waves) was against me. In such cases, I was forced to employ techniques new to me in making up time. I wrote:

> The canoe usually moves at about 4 km per hour, but I sensed that I was making about 2 km per hour under less-than-ideal conditions. As a result, I was forced to employ a sit-and-switch method to paddling and felt I had to canoe longer during the day to make up for lost time.

I was also aware of the temporally bounded nature of the canoe trip. This mindset compelled me to think about the trip through Clawson and Knetsch's (1978) five-stage outdoor recreation model: anticipation, travel to the site, on-site experience, travel home and recollection. Yes, the on-site experience is especially important, but the other four stages also have meaning to me.

> Canoe trips with students in northern Saskatchewan were a big part of my early professional career. Each year we would finish the trip and visit the same Family Restaurant in Prince Albert on the way home for a BIG lunch. We occupied some hybrid or transitional state between wilderness (dirty tripper) and civilised members of society as we changed into our clean clothes left behind in our vehicles.

Reference to Health and Safety was consistently front of mind. Most passages focused on psychological health, the therapeutic value of canoe tripping, and mental (and physical) renewal. For me, however, there was a strong focus on Health and Safety because travelling solo in a wilderness setting demands it. For example, I was keenly focused on eating well regarding calorie intake and the type of calories and taking supplements to keep my energy levels up. Health and Safety emerged as one of the most critical themes in the canoe trip, which I viewed as a carryover from running canoe trips with students where medical assistance is far away. Safety included, for example, eating slowly, where and how to paddle in windy conditions, and choice of campsite:

> For many years, I have made it a practice to camp on islands for solitude because islands are typically in the middle of the lake and have more wind to carry away the bugs (mosquitoes, black flies and deer flies) and I am less likely to encounter bears.

Finally, this was the first time I have taken a GPS communication device on a canoe trip. These communication devices allow one to send a pre-written email. However, they also communicate with your smartphone, so I found myself sending quick emails to my partner, which, although comforting, took something away from the authentic wilderness experience.

Discussion

Figure 13.1 shows that deep peace is a function of many variables identified through the autoethnographic approach employed in this study. My task in the coming pages is to make sense of these dimensions by shedding light on different intensities of peace – no peace, shallow peace, intermediate peace, and deep peace. Deep peace, I argue, represents the pinnacle both within each of the eight dimensions and as a culmination of such.

There was not a single situation in which I felt a sense of *no peace* during the trip. Part of this is likely a function of my experience planning and executing canoe trips over many years – learning about what I need, and do not need, to take on canoe trips. There was also considerable thought placed into this trip, given the privilege of time that I had over many months. I had good weather throughout and the opportunity to choose good campsites that met my needs. However, I have experienced situations and trips that reflect no peace. Situations included tents that leaked, insufficient clothing, partners not on the same page with me in terms of gear, food, and decision-making, getting lost, too many bugs, and transportation problems, all of which presented conditions that compromised peace.

The two domains that I categorised as *shallow peace* in Figure 13.1 were (a) Responsibility and (b) Space, Place and Time. Responsibility in action (i.e. being responsible and sustainable through ethical purchasing and protecting the natural world) emerged as an important theme. Weeden (2011) argues that a heightened sense of inner peace is gained in supporting local communities and protecting the natural world. Lee (2011) found that place attachment, recreation participation, and conservation-mindedness positively impact environmentally responsible behaviour. I confirmed such in my experience because of my history with Temagami and the importance of this region from a preservation standpoint. Home and family responsibilities continued to occupy my thoughts as I left my partner to take care of the house and dogs and drive children to various events. I believe, however, that much of this concern with home was unfounded because, as a former canoeist herself, my partner is supportive of these trips, so it is me, more than her, that feels the burden of absence.

Space, Place and Time were collapsed into one category, but future studies may wish to place Time into its own category because there was a great deal of variability to consider for all three of these factors together. Time was the great agitator of this trip. Unlike Responsibility, I felt that taking more time away from family and home was to burden others with my responsibilities. However, as already noted, in one sense, I was not able to slow down and sink deeply into the splendour that is Temagami. In another sense, I was being true to myself. Resting at the back end of the trip with little distance to cover has always been an approach I have taken in canoe tripping. Notwithstanding, I feel as though I was unable to fully

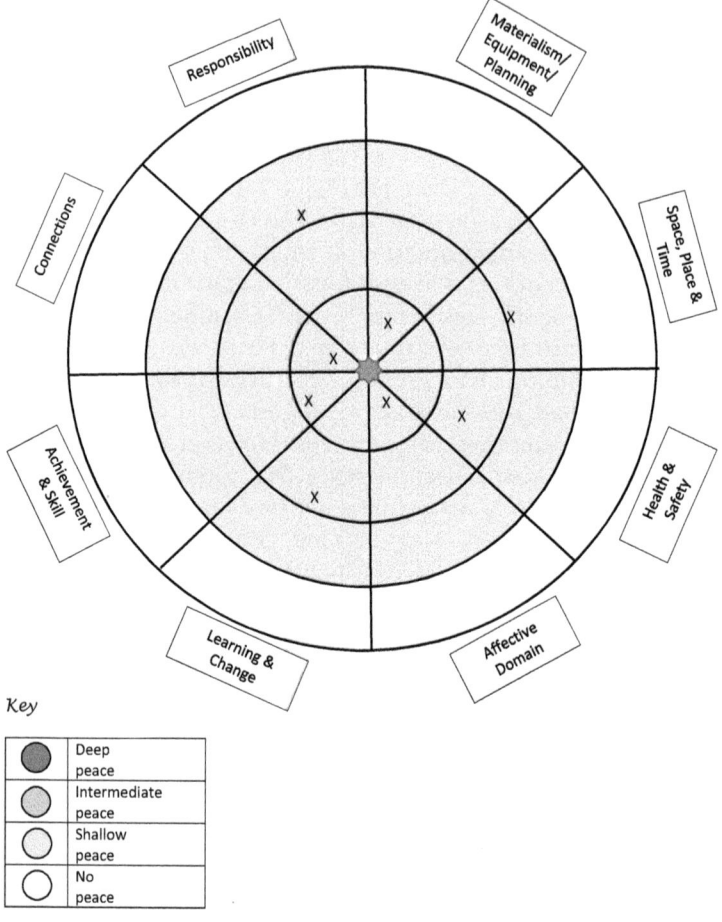

Figure 13.1 Peace themes

achieve a state of mindfulness through slow tourism by deeply immersing myself in the moment and setting (see Farkić et al., 2020)

I experienced *intermediate peace* in two domains. The first, Learning and Change, occupies the outer realm of the intermediate peace circle in Figure 13.1, because I was not happy with my current level of environmental knowledge. I felt as though I should be doing more to better understand the ecology of the Temagami region. One of the core elements of the ecotourism experience is getting into nature and learning as much about these places as possible. I concur with James (1981) on how canoe trips can allow for a reorientation of values as we get a better glimpse into the interior of the self and the need and willingness for change.

Another area of change was felt in my approach to negotiating the more challenging aspects of the trip. The results indicate that a big focus of this change was on not needing to be a 'hero' (Robins, 1943) as a function of age, fitness and perhaps maturity. Following this train of thought, Health and Safety were categorised as intermediate peace because of the need to maintain both on a solo wilderness trip and the common preoccupation with making sure that I was taking care not to step over my physical boundaries – that is, pushing too hard – but also challenging myself to an acceptable degree. Rantala and Varley (2019) contrast the lightness and heaviness of tourist wild camping, with lightness referring to letting go of urban encumbrances and heaviness refer to both heavy packs and the responsibility of ensuring that participants are safe in wilderness spaces. Both of these states were evident in my trip. Measured physical effort, therefore, played a decisive role in my actions and behaviours throughout the trip. The necessity of taking newer technology provided a degree of cognitive dissonance about old school approaches to canoe tripping and expectations around safety given the fact that high-tech companies are well known for pushing such phrases through marketing the use of such devices (Nagle & Vidon, 2021).

I felt a sense of *deep peace* in four of the eight domains in Figure 13.1: Achievement and Skill; Connections; Materialism, Equipment and Planning; and Affective States. Achievement resonates well with the thoughts of Brébeuf (Kenton, 1954, cited in Franks, 1977) and Franks. While suffering for God led to deep peace for Brébeuf, for Franks (1977) the recreational canoeist's suffering and fatigue have more to do with the existential reasons for participation, especially on a solo excursion. In this regard, psychological capital (Luthans *et al.*, 2007) and existential capital (Nettleton, 2013) emerged as more important. Psychological capital refers to having the confidence to undertake challenging tasks, optimism about succeeding at these tasks, perseverance towards the satisfaction of goals and resilience to bounce back when presented with adversity. Existential capital provides a theoretical explanation for Brébeuf's and Franks's deep peace experiences. I share (and felt) the weight of the journey punctuated by several moments of physical suffering.

The preceding mindset fits well with emerging tourism studies research on the need to explore subjective experiences beyond hedonic enjoyment; that is, hardship can be a component of eudaimonia (Knobloch *et al.*, 2016). In addition, evidence points to the importance of wilderness canoe trips for older adults as a 'gentle "push back" against the vicissitudes of ageing, and perhaps even as a metaphor for a mental and physical reawakening' (Cameron & Grau, 2005: 8), with which I emphatically concur.

The aspect of the trip that I felt most at peace with was the theme of Connections. As I move back through the literature and compare it against my trip, I see the canoe trip as providing an opportunity to connect with the self, other people and history (Benidickson, 1997). Despite the rich base of literature on the spiritual nature of the canoe trip, this was not a strong Connections feature for me. However, as noted earlier, what was necessary was the sense of pilgrimage (James, 1981; Vistad *et al.*, 2020) – going back to similar lakes and rivers of my past and metaphysically 'being' with those who made these trips so crucial in my life. The deepness of the experience was sustained through memories of these trips as I sat on smooth Canadian Shield rocks at sunset and listened to the sounds of nature and past voices.

The Materialism, Equipment and Planning theme maintains a strong presence in theory and practice as recreationists fuss over food and gear, experiment with gear and choose to approach the natural world from materialistic and post-materialistic standpoints. I want to say *pre-materialistic* tendencies over post-materialistic because of the steadfastness of maintaining a more traditional approach to canoe tripping by embracing only simple material affordances and through simple living (as part of the hero persona).

Finally, several Affective States were noted on the trip. Few of these mentions were of a negative tone, such as loneliness. As one who prefers quiet and being alone, the experience was representative of my personality. Those who choose to engage in such a trip who are more gregarious and prone to loneliness will undoubtedly encounter a heightened level of cognitive dissonance. Positive feelings and states also resonate with the connecting to self sub-theme under Connections, but I deemed it essential to separate these feelings from the inner journey of the self that is representative of the latter category. Feelings of empowerment, awareness, inspiration, the privilege to be in Temagami and experiencing pleasure over pain expanded on the psychological aspects of solo canoeing in general and my trip more specifically.

Conclusion

The primary purpose of this study was to identify the fundamental characteristics of deep peace based on a one-week solo wilderness canoe trip to Temagami, Canada. The results indicate that peace is not a homogeneous state but rather one of degree. The final objective of this journey is to 'extend the trip' one step further by synthesising the results into a definition of deep peace:

> A multidimensional state of being based on one's immersion into an activity that leads to a profound sense of inner harmony, wisdom, and contentment.

The sheer number of different states experienced through the trip suggests that deep peace is *multidimensional*. State of being is essential to include over state of mind in this definition because it suggests not just a focus on mind but rather mind and body.

The second aspect of the definition, *immersion*, is an essential facet of deep peace according to the depth and intensity of the experience. I argue that deep peace can be achieved through a momentary experience such as the completion of a long and challenging portage (a micro-level experience), a long and challenging day of paddling (a meso-level experience or episode), or indeed the whole trip itself (a macro-level experience). However, while deep peace occurs in pockets of time and space, I argue that achieving deep peace over the entire duration of the TOR experience should be the ultimate goal.

The third feature of the definition of deep peace is *activity*. The experience of deep peace in solo canoeing may be different for different canoeists, may be different under different settings and circumstances, and is undoubtedly different between TOR activities. For example, the adventure traveller may find more profound peace in the Achievement and Skill domain. By contrast, the ecotourist may find deeper peace in the Learning and Change, Connections (to nature), and Responsibility domains.

Harmony is viewed as synchronisation or congruence between an individual and the natural world under the belief that such cannot be achieved in the absence of both (Horwood, 1999). *Inner wisdom* (Phipps, 1987) references awareness of oneself by being attuned to our feelings, sensations, and intuitions. The solo trip may be good for the soul because it induces a peeling back of the layers of self in the absence of others to redirect thoughts and actions. Facing the self may be a blessing, but it might also be a challenge as we discover who we are at the very core. Whether we find a true sense of authentic happiness or dread and despair is of considerable importance. The final main feature of the definition, *contentment*, directly correlates to peace (Røgild-Müller & Robinson, 2020). However, we can be content under many different circumstances. Such includes preservation or destruction of landscapes, support and care or shameless cheating of others. Important here is a conception of the good that comes from contentment or happiness that occurs not at the expense of others. Such is why inner wisdom (awareness) and harmony are needed alongside contentment. We can be content through inner wisdom (awareness of ourselves and the settings in which we participate) and harmony (acting in a manner that protects the natural world).

Deep peace should be viewed as a metatheoretical device that can act as connective tissue for several other theoretical constructs that currently exist in the TOR literature. Synergies can be built, for example, with serious leisure, specialisation, encounter norms, environmental ethics and values, flow, optimal arousal, mindfulness, SWB and different stages of travel models. Future studies may wish to determine if more experienced canoe trippers can achieve deeper levels of peace given their experience. Additionally, research should investigate if trips to unfamiliar canoe destinations play a role in heightening or diminishing deep peace, or even if a deeper level of peace is actively sought in such experiences.

References

Adler, V.H. and Kwon, S. (2002) Social capital: Prospects for a new concept. *Academy of Management Review* 27, 17–40.

Andersen, F. (2006) I skovensdybestille ro. In C. Andersen, M. Bojesen, J. Bonderup, O. Brunsbjerg, J. Carlsen, H. Dammeyer, R.A. Jensen and K.H. Munk (eds) *Højskolesangbogen* (18th edn) (no. 572). Copenhagen: Edition Wilhelm Hansen.

Anon. (2016) Solo act. *Family Camping & Canoeroots*. Spring, 8.

Badè, W.F. (ed.) (1916) Introduction. In J. Muir, *A Thousand-Mile Walk to the Gulf*, edited by W.F. Badè (pp. ix–xxvi). Boston: Houghton Mifflin Co.

Beckford, C.L., Jacobs, C., Williams, N. and Nahdee, R. (2010) Aboriginal environmental wisdom, stewardship, and sustainability: Lessons from the Walpole Island First Nations, Ontario, Canada. *The Journal of Environmental Education* 41 (4), 239–248. https://doi.org/10.1080/00958961003676314.

Benidickson, J. (1997) *Idleness, Water, and a Canoe: Reflections on Paddling for Pleasure*. Toronto: University of Toronto Press.

Bishop, S.R., Lau, M., Shapiro, S., Carlson, L., Anderson, N.D., Carmody, J., Segal, Z.V., Abbey, S., Speca, M., Velting, D. and Devins, G. (2004) Mindfulness: A proposed operational definition. *Clinical Psychology: Science and Practice* 11, 230–241.

Bourdieu, P. (1977) *Outline of a Theory of Practice*. Cambridge: Cambridge University Press.

Butz, D. (2010) Autoethnography as sensibility. In D. DeLyser, S. Aitken, S. Herbert, M. Crang and L. McDowell (eds) *The SAGE Handbook of Qualitative Research in Human Geography* (pp. 138–155). London: SAGE.

Butz, D. and Besio, K. (2009) Autoethnography. *Geography Compass* 3 (5), 1660–1674.

Caduto, M.J. (1998) Viewpoint: Ecological education: A system rooted in diversity. *The Journal of Environmental Education* 29 (4), 11–16. https://doi.org/10.1080/00958969809599123.

Cameron, J. and Grau, M. (2005) Understanding the wilderness canoe trip experience of older adults. *Pathways: The Ontario Journal of Outdoor Education* 17 (1), 4–8.

Chakraborty, A. (2019) Does nature matter? Arguing for a biophysical turn in the ecotourism narrative. *Journal of Ecotourism* 18 (3), 243–260. https://doi.org/10.1080/14724049.2019.1584201.

Chirozva, C. (2015) The Great Limpopo Transfrontier Conservation Area includes community agency and entrepreneurship in ecotourism planning and development. *Journal of Ecotourism* 14 (2–3), 185–203. https://doi.org/10.1080/14724049.2015.1041967.

Clawson, M. and Knetsch, J.L. (1978) *Economics of Outdoor Recreation* (4th edn). Baltimore: The Johns Hopkins University Press.

Cooper, R. and Lilyea, B.V. (2022) I'm interested in autoethnography, but how do I do it? *The Qualitative Report* 27 (1), 197–208. https://doi.org/10.46743/2160-3715/2022.5288.

Csikszentmihalyi, M. (1990) *Flow: The Psychology of Optimal Experience*. New York: HarperCollins.

Curtin, S. (2005) Nature, wild animals and tourism: An experiential view. *Journal of Ecotourism* 4 (1), 1–15. https://doi.org/10.1080/14724040508668434.

D'Amato, L.G. and Krasny, M.E. (2011) Outdoor adventure education: Applying transformative learning theory to understanding instrumental learning and personal growth in environmental education. *The Journal of Environmental Education* 42 (4), 237–254. https://doi.org/10.1080/00958964.2011.581313.

Diener, E., Suh, E.M., Lucas, R.E. and Smith, H.L. (1999) Subjective wellbeing: Three decades of progress. *Psychological Bulletin* 125 (2), 276–302.

Dunkin, J. (2019) *Canoe and Canvas: Life at the Encampments of the American Canoe Association, 1880–1910*. Toronto: University of Toronto Press.

Eliason, S.L. (2016) Access to public resources on private property: Resident hunter perceptions of the commercialisation of wildlife in Montana. *Journal of Outdoor Recreation and Tourism* 16, 37–43.

Ellis, C. and Bochner, A.P. (2000) Autoethnography, personal narrative, reflexivity: Researcher as subject. In N.K. Denzin and Y.S. Lincoln (eds) *Handbook of Qualitative Research* (2nd edn) (pp. 733–768). Thousand Oaks: SAGE.

Erickson, B. (2008) Canoe nation: Canoes and the shifting production of space through white Canadian masculinities. *Antipode* 40 (1), 182–184.

Erickson, B. and Wylie Krotz, S. (2021) *The Politics of the Canoe*. Winnipeg: University of Manitoba Press.

Farkić, J., Filep, S. and Taylor, S. (2020) Shaping tourists' wellbeing through guided slow adventures. *Journal of Sustainable Tourism* 28 (12), 2064–2080. https://doi.org/10.1080/09669582.2020.1789156.

Finlay, L. (1998) Reflexivity: An essential component for all research? *British Journal of Occupational Therapy* 61 (10), 453–456.

Fletcher, R. (2014) *Romancing the Wild: Cultural Dimensions of Ecotourism*. Durham, NC: Duke University Press.

Franks, C.E.S. (1977) *The Canoe and White Water: From Essential to Sport*. Toronto: University of Toronto Press.

Gibbs, G. (2008) *Analyzing Qualitative Data*. London: SAGE.

Grimwood, B.S.R. (2011) 'Thinking outside the gunnels': Considering natures and the moral terrains of recreational canoe travel. *Leisure/Loisir* 35 (1), 49–69. https://doi.org/10.1080/14927713.2011.549196.

Grimwood, B.S.R. (2015) Advancing tourism's moral morphology: Relational metaphors for just and sustainable arctic tourism. *Tourist Studies* 15 (1), 3–26.

Guasca, M., Vanneste, D. and Van Broeck, A.M. (2022) Peacebuilding and post-conflict tourism: Addressing structural violence in Colombia. *Journal of Sustainable Tourism* 30 (2–3), 427–443. https://doi.org/10.1080/09669582.2020.1869242.

Haun-Moss, B. (2002) Layered hegemonies: The origins of recreational canoeing desire in the Province of Ontario. *Topia: Canadian Journal of Cultural Studies* 7, 39–55.

Henderson, B. (1999) The canoe as a way to another story. In J. Jennings, B. Hodgins and D. Small (eds) *The Canoe in Canadian Cultures* (pp. 183–198). Winnipeg: Natural Heritage Books.

Higgins-Desbiolles, F., Blanchard, L.-A. and Urbain, Y. (2022) Peace through tourism: Critical reflections on the intersections between peace, justice, sustainable development and tourism. *Journal of Sustainable Tourism* 30 (2–3), 335–351. https://doi.org/10.1080/09669582.2021.1952420.

Hitchner, S., Schelhas, J., Brosius, J.P. and Nibbelink, N. (2019) Thru-hiking the John Muir Trail as a modern pilgrimage: Implications for natural resource management. *Journal of Ecotourism* 18 (1), 82–99. https://doi.org/10.1080/14724049.2018.1434184.

Hodgins, B. (1988) Canoe irony: Symbol and harbinger. In J. Raffan and B. Horwood (eds) *Canexus: The Canoe in Canadian Culture* (pp. 45–57). Toronto: Betelguese Books.

Horwood, B. (1999) The Dao of paddling. In J. Jennings, B.W. Hodgins and D. Small (eds) *The Canoe in Canadian Cultures* (pp. 62–73). Ottawa: Natural Heritage Books.

Hoyle, G. (1999) The dark side of the canoe. In J. Jennings, B.W. Hodgins and D. Small (eds) *The Canoe in Canadian Cultures* (pp. 212–223). Ottawa: Natural Heritage Books.

Ingolfsdottir, A. and Gunnarsdottir, G. (2020) Tourism as a tool for nature conservation? Conflicting interests between renewable energy projects and wilderness protection in Iceland. *Journal of Outdoor Recreation and Tourism* 29, 100276.

James, W. (1981) The canoe trip as religious quest. *Studies in Religion/Sciences Religieuses* 10 (2), 151–166.

Janesick, V.J. (2010) *Oral History for the Qualitative Researcher: Choreographing the Story*. New York: Guilford Press.

Jirásek, I. and Hurych, E. (2019) Experience of long-term transoceanic sailing: Cape Horn example. *Journal of Outdoor Recreation* 28, 1–9.

Kelly, J. (1987) *Freedom to Be: A New Sociology of Difference*. New York: Macmillan.

Kennedy, S., MacPhail, A. and Varley, P.J. (2019) Expedition (auto)ethnography: An adventurer-researcher's journey. *Journal of Adventure Education and Outdoor Learning* 19 (3), 187–201. https://doi.org/10.1080/14729679.2018.1451757.

Kenton, E. (1954) *The Jesuit Relations and Allied Documents*. New York: Vanguard.

Kim, H., Lee, S., Uysal, M., Lim, J. and Ahn, K. (2015) Nature-based tourism: Motivation and subjective well-being. *Journal of Travel & Tourism Marketing*, 1–21. https://doi.org/10.1080/10548408.2014.997958.

Knobloch, U., Roberston, K. and Aitken, R. (2016) Experience, emotion and eudaimonia: A consideration of tourist experiences and well-being. *Journal of Travel Research* 56 (5), 651–622. https://doi.org/10.1177/0047287516650937.

Knowles, J.G. (1992) Geopiety, the concept of sacred place: Reflections on an outdoor educational experience. *Journal of Experimental Education* 15 (1), 6–12.

Kowalewski, D. (2002) Teaching deep ecology: A student assessment. *The Journal of Environmental Education* 33 (4), 20–27. https://doi.org/10.1080/00958960209599150.

Lai, Po-Hsin and Shafer, S. (2005) Marketing ecotourism through the internet: An evaluation of selected ecolodges in Latin America and the Caribbean. *Journal of Ecotourism* 4 (3), 143–160. https://doi.org/10.1080/JET.v4.i3.pg143.

Lee, T.H. (2011) How recreation involvement, place attachment and conservation commitment affect environmentally responsible behavior. *Journal of Sustainable Tourism* 19 (7), 895–915. https://doi.org/10.1080/09669582.2011.570345.

Lepp, A. and Herpy, D. (2015) Paddlers' level of specialisation, motivations and preferences for river management practices. *Journal of Outdoor Recreation and Tourism* 12, 64–70.

Lewis, M.S., Lime, D.W. and Anderson, D.H. (1996) Paddle canoeists' encounter norms in Minnesota's boundary waters canoe area wilderness. *Leisure Sciences* 18 (2), 143–160. https://doi.org/10.1080/01490409609513278.

Lindholst, A.C., Caspersen, O.H. and Konijnendijk Van Den Bosch, C.C. (2015) Methods for mapping recreational and social values in urban green spaces in the Nordic countries and their comparative merits for urban planning. *Journal of Outdoor Recreation and Tourism* 12, 71–81. https://doi.org/10.1016/j.jort.2015.11.007.

Luthans, F., Youssef, C.M. and Avolio, B.J. (2007) *Psychological Capital: Developing the Human Competitive Edge*. New York: Oxford University Press.

Maina-Okori, N.M., Koushik, J.R. and Wilson, A. (2018) Reimagining intersectionality in environmental and sustainability education: A critical literature review. *The Journal of Environmental Education* 49 (4), 286–296.

McCabe, S. and Johnson, S. (2013) The happiness factor in tourism: Subjective wellbeing and social tourism. *Annals of Tourism Research* 41, 42–65.

McDermott, L. (2004) Exploring intersections of physicality and female-only canoeing experiences. *Leisure Studies* 23 (3), 283–301. https://doi.org/10.1080/0261436042000253039.

Moscardo, G. (2009) Understanding tourist experience through mindfulness theory. In M. Kozak and A. Decrop (eds) *Handbook of Tourist Behavior* (pp. 99–115). New York: Routledge.

Moufakkir, O. and Kelly, I. (2010) *Tourism, Progress and Peace*. Wallingford: CABI.

Mueller Worster, A. (2017) Sustainability frontiers: Critical and transformative voices from the borderlands of sustainability education. *The Journal of Environmental Education* 48 (2), 139–141. https://doi.org/10.1080/00958964.2016.1200526.

Mullins, P.M. (2009) Living stories of the landscape: Perception of place through canoeing in Canada's North. *Tourism Geographies* 11 (2), 233–255. https://doi.org/10.1080/14616680902827191.

Nagle, D.S. and Vidon, E.S. (2021) Purchasing protection: Outdoor companies and the authentication of technology use in nature-based tourism. *Journal of Sustainable Tourism* 29 (8), 1253–1269. https://doi.org/10.1080/09669582.2020.1828432.

Nettleton, S. (2013) Cementing relations within a sporting field: Fell running in the English Lake District and the acquisition of existential capital. *Cultural Sociology* 7 (2), 196–210.

Newbery, L. (2003) Will any/body carry the canoe? A geography of the body, ability, and gender. *Canadian Journal of Environmental Education* 8, 204–216.

O'Leary, B.S., Lingholm, M.L., Whitford, R.W. and Freeman, S.E. (2002) Selecting the best and the brightest: Leveraging human capital. *Human Resource Management* 41, 325–340.

Phipps, J.-F. (1987) Deep peace. *The Trumpeter* 4 (2), 32–34.

Pratt, S. and Liu, A. (2016) Does tourism really lead to peace? A global view. *International Journal of Tourism Research* 18, 82–90.

Raffan, J. (1999a) *Bark, Skin and Cedar: Exploring the Canoe in Canadian Experience*. Toronto: HarperCollins.

Raffan, J. (1999b) Being there: Bill Mason and the Canadian canoeing tradition. In J. Jennings, B.W. Hodgins and D. Small (eds) *The Canoe in Canadian Cultures* (pp. 15–27). Ottawa: Natural Heritage Books.

Rantala, O. and Varley, P. (2019) Wild camping and the weight of tourism. *Tourist Studies* 19 (3), 295–312.

Reed-Danahay, D. (1997) Introduction. In D. Reed-Danahay (ed.) *Auto/ethnography: Rewriting the Self and the Social* (pp. 1–20). London: Routledge.

Robins, J.D. (1943) *The Incomplete Anglers*. Toronto: Collins.

Røgild-Müller, L. and Robinson, J. (2020) Emergence and experience of 'peace of mind': What can classic writers tell us? *Human Arenas* 5, 298–311. https://doi.org/10.1007/s42087-020-00147-1.

Ross, A. (1995) *Coke Stop in Emo: Adventures of a Long-Distance Paddler*. Toronto: Key Porter Books.

Rutter, J. (1978) *A Gaelic Blessing*. Hinshaw Music. Oxford: Oxford University Press.

Sawchuk, W. (2016) Riding the divide – balancing resource extraction and conservation in the Muskwa-Kechika region of northern British Columbia, Canada. *Journal of Ecotourism* 15 (3), 285–293. https://doi.org/10.1080/14724049.2016.1189557.

Thomas, A. (1999) Paddling voices: There's the poet, voyager, adventurer and explorer in all of us. In J. Jennings, B.W. Hodgins and D. Small (eds) *The Canoe in Canadian Cultures* (pp. 175–182). Ottawa: Natural Heritage Books.

Tsevreni, I. (2021) Nature journalling as a holistic pedagogical experience with the more-than-human world. *The Journal of Environmental Education* 52 (1), 14–24. https://doi.org/10.1080/00958964.2020.1724854.

Valsiner, J. (2001) Process structure of semiotic mediation in human development. *Human Development* 44, 88–97. https://doi.org/10.1159/000057048.

Vistad, O.I., Øian, H., Williams, D.R. and Stokowski, P. (2020) Long-distance hikers and their inner journeys: On motives and pilgrimage in Nidaros, Norway. *Journal of Outdoor Recreation and Tourism* 31, 100326. https://doi.org/10.1016/j.jort.2020.100326.

Weeden, C. (2011) Responsible tourist motivation: How valuable is the Schwartz value survey? *Journal of Ecotourism* 10 (3), 214–234. https://doi.org/10.1080/14724049.2011.617448.

Wigglesworth, J. (2018) Writing gendered embodiment into outdoor learning environments: Journaling for critical consciousness. In T. Gray and D. Mitten (eds) *The Palgrave International Handbook of Women and Outdoor Learning* (pp. 789–800). Palgrave Studies in Gender and Education. Cham: Palgrave Macmillan. https://doi.org/10.1007/978-3-319-53550-0_54.

Wilson, H. (2011) *Temagami: A Wilderness Paradise*. Richmond Hill, ON: Firefly Books.
Xu, W., Rodrigues, M.A., Zhang, Q. and Liu, X. (2014) The mediating effect of self-acceptance in the relationship between mindfulness and peace of mind. *Mindfulness* 6 (4), 797–802. https://doi.org/10.1007/s12671-014-0319-x.
Ym, M. (2022) Tourism and peace relationship between tourism, peace & violence and conflict. *Journal of Tourism Studies and Hospitality Research* 3 (1), https://www.pubtexto.com/article_pdf/680/1280/pubtexto_680_1280_25022022031026.pdf.

Conclusion

Richard Butler

Tourism is a simple concept, namely, travelling to a temporary location for a period of rest/relaxation/recreation in order to experience pleasure before returning home, but increasingly, in academic terms at least, it is becoming ever more complex, problematical, and sometimes disappointing. We, the academic community, examine and scrutinise the phenomenon for variations, for motivations, for impacts and for problems, only very rarely studying tourism for its positive role in providing hope, pleasure, and satisfaction. For those engaged commercially in the tourism industry, motivations are primarily economic, such as employment, income, and return on investment, and their immediate concerns are mostly related to survival in the business sense, given the competitive nature of tourism and problems such as the COVID-19 pandemic. For residents of tourist destinations, their emotions regarding tourism and tourists will vary, in part depending on the nature of their engagement with both tourism and tourists; some, particularly those hosting tourists or employed in some capacity in tourism, will share the general industry's views on the economic benefits of tourism, accepting in many cases the lack of an alternative to tourism in terms of employment and income. Other residents, often those gaining no direct economic benefits from tourism, may understandably be negative about the disruption and inconvenience they experience because of the presence and pressures from tourists in their community.

For tourists, however, the situation is somewhat different, for despite the inconveniences, stress, and expense of travelling to and from their destination(s), the other financial costs involved, and various possible difficulties encountered, the holiday is a time of hope and anticipation of pleasure, in whatever form it may be sought (Lohman & De Bloom, 2021). It may be the joy of freedom, from work, from school, from regimentation and the grind of 'normal' life; it may be the joy of experiencing different landscapes, cultures and cuisine, as Das and Roy note; it may be learning new skills; the challenge of new activities or re-experiencing past pleasures; or simply meeting people, both familiar and new. Whatever their intentions and desires, one key characteristic of every tourist on holiday will be hope, namely, that their expectations of pleasure will be met and the recollections of that pleasure will be added to their store of positive memories. As academics, at times we are guilty of forgetting the phenomena of hope, of pleasure, and of enjoyment and their role in tourism. The chapters in this volume have gone some way towards correcting that neglect or forgetfulness as they have analysed, reviewed and described some of the issues involved in the positive side of tourism. Appearing at a time (late 2022) when the COVID-19 pandemic and associated travel restrictions, which halted most tourist activity across the globe, appear to be ending , and many people have come to appreciate it more clearly, this volume's focus on the positive benefits of hope and enjoyment that tourism can bring is even more relevant.

Recollection and Anticipation

In studying tourism there is a tendency to analyse it to increasing depth, often resulting in subdivision and categorisation of motivations, activities, tourism and tourists themselves. Thus, there are over a hundred sub-forms of tourism in the academic literature, rather overlooking the reality that they are all parts of a much larger whole phenomenon – tourism – and that subdivision can sometimes hide more than it reveals. In classifying tourism and its participants as mass tourists, cultural tourists, sustainable tourists, and so forth (ignoring, for the moment, related terms like visitors, guests, vacationers, and, perhaps least meaningful of all, travellers, with its misplaced hint of superiority), what is lost is what is common to all of these individuals, elements such as anticipation, satisfaction, enjoyment, and hope.

Recollections and anticipation are key elements with regards to the pursuit of hope in tourism. They are, to some extent, the starting point of all recreation/leisure/tourism journeys (Clawson & Knetsch, 1966), whereby potential tourists look back to previous holidays and forward to their anticipated holiday, full of expectation for the pleasure they hope it will bring. In most cases such anticipation is related to and dependent on recollections of previous holidays, both positive and negative, and these recollections play a key role in decision-making about a number of variables, such as weather, cost, opportunities (gained and abandoned), alternatives, time involved, and the balance that emerges between different possible destinations (Gartner, 1994). Recollections of poor experiences, such as the journeys to and from the destination; disappointments at the destination of higher than anticipated costs and of misplaced confidence in the availability of attractions; bad, rather than good, weather; and poor group dynamics, all potentially resulting in a low level of satisfaction, can deter tourists from returning to a specific destination, just as strongly positive recollections can induce a person or party to revisit past locations. Familiarity, as Bowen and Clarke discuss in their chapter, can inspire confidence that anticipation will be rewarded and satisfied, based on previous experiences. Such confidence can be seen in the number of people who invest in holiday properties, the summer cottage beloved by Scandinavians (Hall & Müller, 2018) and North Americans (Wolfe, 1977) in particular, or apartments and condominiums in European and North American destinations. In such cases, recollections of a 'second home' are normally positive and also encourage nostalgia and inertia, both powerful if understudied factors influencing tourist destination choice and behaviour. Summer home communities (Coppock, 1977) represent an interesting paradox in tourism, whereby holidaymakers often meet and make friends over the years by returning to the same specific location, reliving experiences from one year to the next, thus defeating what is often thought of as one of the characteristic motivations of tourists, to experience something new and different, the 'pull' of the 'push and pull' set of motivations (Dann, 1981).

Anticipation is not a constant, one of the characteristics of tourism and tourists is dynamism, and while the overall pattern of distribution of tourists and their behaviours may have remained relatively similar over the past half century, at the individual (family, party, group) level, there is often considerable variation year on year. Families age, and what may be remembered as a satisfying destination for a family with young children will not be perceived as such by teenage members of that family later on. Destinations favoured by young singles may be recollected at times with horror as well as embarrassment and guilty pleasure by parents later in life. The elderly may recollect, sometimes poorly, sometimes with great accuracy, sometimes pleasurably and sometimes wistfully, past destinations, while anticipating that relaxation and comfort may be more appropriate at their stage in life, and may abandon thoughts of recreating previous experiences or fantasies. The attempts, as outlined in the two films discussed by Moscardo, to experience new places and recover and find replacements do not always work out well, the fear of which may well explain the attraction of the familiar, particularly in the case of parties with small children.

This established pattern has of course been upturned by the COVID-19 pandemic and the resulting restrictions, lockdowns, and non-availability of various destinations, as Vada and Scott review in their chapter. As discussed elsewhere (Butler, 2021), the threat and requirement of quarantine, for example, has made the journeys to and from destinations much more significant and in some cases more important than the destination actually selected. Spending two weeks in quarantine on arrival and again on returning home makes two weeks at a destination much less attractive, meaning great inconvenience and uncertainty and great additional costs. There can be no recollection involved in such cases, as no one has experienced a global pandemic such as has been experienced over the past two years, and there can be no anticipation based on fact either as almost no one knows what to expect in such a pandemic. Anticipation of the likelihood of delays, restrictions, quarantines and outright bans imposed during the holiday itself have not been encountered by many potential tourists, except those who have suffered problems such as conflicts and natural disasters occurring while on holiday. Potential tourists have had to rely on official, often rapidly changing announcements on restrictions, and, inevitably, on the dubious value of social media reports (Gretzel, 2019), notorious for their factual errors, biases, and misinformation, or fake news. Travel during the COVID-19 pandemic has come to resemble the scenario framed by Robert Louis Stevenson (1878: 2) when he wrote, 'To travel hopefully is a better thing than to arrive'. Such a situation has resulted in the general irrelevance of recollection and anticipation and a greater reliance on hope, sometimes despite reality, for obtaining pleasure from any holiday in 2020 and 2021.

Pleasure and Satisfaction

While it may be accepted that pleasure is the basic motivation for taking a holiday, the way that pleasure, and thus satisfaction, is obtained takes a great variety of forms. The differences between hedonism and eudaimonism have been discussed in several of the chapters in this volume (particularly by Kozak, Vada and Scott, and implicitly by Polus and Carr) and it is difficult not to view such different motivations through a personal lens. Many academics seem to do this in their dislike of mass tourism, perhaps the personification of hedonistic tourism, often designed to produce the maximum level of instant personal pleasure, sometimes regardless of consequences. Such a dislike is echoed by many travel writers, particularly in the 'respectable' media, where passing negative comments on 'the masses' and expressing preferences for unspoilt (normally meaning where tourists are absent) destinations available to the fortunate affluent few and freeloaders like many of the writers themselves. Such emotions are not new; they were expressed in the mid-19th century by John Ruskin, who, on visiting Venice, found tourists there who did not share his intense admiration and knowledge of architecture and viewed their presence as not only annoying but problematic, complaining bitterly. Many with similar views express the same emotion today with respect to Venice and other locations experiencing overtourism (Visentin & Bertocchi, 2019), ignoring or missing the fact that many more people experience pleasure from visiting such places despite their own 'ignorance' (as Ruskin would have put it) and the numbers of tourists present. Wheeller (2006) reviewed and revealed the inconsistency of the few deploring the pleasurable experiences of the many, just as Butcher (2020) has done more recently.

Pleasure comes in many forms and the old adage that 'one man's meat is another man's poison' is equally true in tourism. The attractions of Benidorm, Magaluf, Ibiza and Ayia Napa may not provide pleasure to some, but they do to many. It should not be appropriate to rule on anyone else's pleasure (as long as it is legal and free from harm to others), except perhaps when those forms of pleasure impact negatively on others and their quality of life. Here tourism is on dangerous ground, as the spectre of overtourism is based very much on the dissatisfaction of residents with the presence and activities of excessive numbers of tourists (Mihalic, 2020). The number of visitors needs to be considered apart from the activities of visitors, as they are two distinct parts of the same problem of overtourism. Excessive numbers of visitors are often the result of the destination promoting itself beyond the appropriate level as far as residents are concerned, and thus the misnamed 'managers' of tourism are themselves at least partly to blame. The behaviour of those visitors is a much more difficult topic to deal with, as what may be pleasure to them can cause intense annoyance and even danger to residents and others. To some extent tourists will self-segregate; those seeking hedonistic experiences whatever the cost to others will not be accompanied by fellow tourists to the same destination, although they may share the same aeroplane, train, boat, or motorway to get to and from the destination. Increasingly, however, destinations are becoming more differentiated on social media and in their own promotion, so the problem may be somewhat self-solving.

Those who seek pleasure and satisfaction in what have been called 'alternative' forms of tourism would appear to be increasing in number, although all tourism, until 2020, has been increasing in numbers, so claims of 'fastest growing' mean little if the starting point is extremely low. The rise in 'meaningful experiences' is undoubtedly reality, and the motivations for gaining pleasure, for example from volunteering and helping others, as reviewed by Polus and Carr, are complex and require more detailed knowledge of how and where tourism fits in people's search for meaning, as discussed by Fraga and Borges. The basis of reciprocal altruism is an academic context probably unknown to many tourists, although it may be particularly applicable to those engaging in volunteer tourism. Tourism research has been limited in the field of psychology, as Fraga and Borges and Kozak note, and thus misconceptions of terminology and concepts is not uncommon, as Moscardo points out in the context of mindfulness.

Contradictions and Inconsistencies

In the search for happiness and enjoyment in their holidays, it is clear that tourists take a variety of actions to achieve these goals. Some seek escape and change to attain pleasure, engaging in this search both spatially and internally. The move from the urban to the rural landscape as described by Das and Roy is one of the most common forms of both short and long leisure journeys, sometime lasting for a few hours and in other cases for weeks, allowing for the sampling of different cultures, different visual surroundings, different food and drink, a different pace of life, and the chance to meet with different viewpoints and outlooks. In another case, we can see the desire to use holiday travel to change viewpoints and encourage the emergence of peace and the disappearance of hostility, as Isaac describes as a potential result of tourism and travel. In other cases the escape may be to leave a situation of despair to pursue hope and pleasure and a new beginning, as Moscardo describes, and, even if the result is not always what was desired, it may represent an improvement on the previous situation. Perhaps the most vivid example of escape is that of the Trinidad Carnival described by Coomansingh, where a whole society escapes the regulations of the norm and rejects order for an intense period of joy and hedonism, often sharing this period with visitors who come to partake in the festival. At the other extreme, perhaps, we have the example provided by Fagence, an escape from the present to visit a dark site of the past, only to emerge with feelings of respect and admiration and pleasure at the transformation of his anticipated feelings.

On the other hand, we have those who travel to seek out comfort through familiarity and perhaps an escape from the unknown and outside influences, as demonstrated by those examined by Bowen and Clarke, who gain relaxation, peace of mind, and enjoyment from known and loved settings and companions. A similar sense emerges also from the family-focused vacations studied by Mirehie and Sharayevska, and those readers with children will be familiar with the sentiments discussed and the motivations behind such holidays. The ideas of self-improvement in the family context and the maintenance of traditions expressed there are closely related to efforts to recover from the travails of the COVID-19 pandemic, as shown by Vada and Scott, with the argument that restoration of well-being, physical and mental, can be achieved through appropriate travel in a domestic context. A similar tone runs through Smith's chapter, where the concept of the retreat, from a stressful to a peaceful atmosphere, echoes Isaac's desire for the attainment of harmony and the personal fulfillment discussed by Kozak and Fraga and Borges in the context of positive psychology.

Several of the chapters comment on the importance of, and focus on, the particularly personal aspect of hope and engagement in tourism for joy and satisfaction. Fagence's chapter is perhaps the most personal, with his recounting of changes in personal outlook and emotions through visiting a specific location, and Smith notes the changed feelings, both physically and emotionally, from deeply personal experiences, albeit in the company of others. A common theme throughout many chapters is that of hope of improvement, both of oneself, and, in Isaac's example, of the arrangement of society. Whether the volunteer tourists studied by Polus and Carr are concerned with self-improvement from their acts of altruism is hard to determine, although it is difficult to imagine that they do not return from holiday with some feeling of satisfaction from helping others and perhaps feel the better for that, and thus improved in some way, rather as Moscardo notes about considering the needs of others. Whether that is their motivation for engaging in volunteering is much more difficult to determine, as we need to know much more about the way that the brain works, as Fraga and Borges argue strongly. Travelling for the purpose of self-improvement is also discussed by Das and Roy in the context of improving one's health and values, and such objectives are implicit in the family holidays and the trips of those striving to overcome the negative effects of COVID-19. Perhaps the only chapter that does not feature this theme is that of Coomansingh, where the carnival experience is aimed at what we might view as pure pleasure, hedonism, but within a social group setting, with no common goals other than freedom from order and the pursuit of joy.

Conclusion

Perhaps the chapters in this volume confuse our understandings of the role and purpose of tourism in the pursuit of happiness and a more positive outlook on life because of the considerable variety of approaches to attaining those goals. Perhaps they serve to emphasise the arguments of Fraga and Borges about how little we really understand the working of the human brain and the need for much more study of neurology and related fields. On the other hand, this volume can be taken as giving a strong impression of the efforts people go to in order to gain pleasure and the great variety of ways in which they find it through tourism. That is, perhaps, the most important message that can be sent; to quote Stevenson (2009/1877: 7) again, 'There is no duty we so much underrate as the duty of being happy'. It is the basic motivation of tourism and related leisure travel and its importance has long been denigrated, criticised, and misunderstood.

While the so-called Puritan work ethic, if it ever existed, seemed focused on working in this life to enjoy leisure in the life hereafter, and to uphold the belief that idleness and pleasure were, if not sinful, at least not to be admired, desired, or pursued, the modern age, with its apparently continuously growing affluence, has allowed ever increasing numbers of people to engage in tourism, in travel for pleasure, with that pleasure being found in both great effort and exertion and in relaxation and idleness. Those who are fortunate enough work more to be able to engage in leisure than they engage in leisure in order to become fitter for work. We ironically comment 'Don't work too hard' or 'Take it easy', phrases that might have been viewed as heretical centuries ago. Re-creation is no longer seen as the purpose of time away from work but as a focus and means in itself. Pleasure, hope, enjoyment, and the pursuit of a higher quality of life are now seen as appropriate goals and objectives. Satisfaction is being viewed as more than a certain level of income, and satisfaction with tourism at the destination level is being seen as being much more than jobs and expenditure generated (Dwyer, 2022); one day it might be measured in happiness provided or pleasure generated, although we are still far from being able to measure such intangibles convincingly.

There is still a reluctance in society and in academia to take the study of hope, joy, pleasure, and, by implication, tourism as seriously as these topics deserve. Tourism as an academic subject of study developed late and is still viewed as of peripheral academic value at best. Its gradual movement from other disciplines into schools of business and management is perhaps indicative of the overimportance given to its ability to attract fee-paying students and its economic significance, to the detriment of its other benefits such as those illustrated in this volume. One effect of the COVID-19 pandemic may be to awaken many to how important tourism (and leisure and recreation) are to a majority of people and how important those elements are to human well-being, both mentally and physically. The economic losses occasioned by the cessation of tourism and other leisure activities have demonstrated how significant such travel is to world and national economies, as well as to small communities and individuals. The disruption to people's lives, particularly by the limitations on or prevention of their pursuit of happiness, has provoked great anger and dissatisfaction, somewhat increased by the inconsistencies of various policies and restrictions. In the current aftermath of the pandemic it would now appear likely that tourism will soon increase back towards its 2019 level, but perhaps with a greater appreciation of what has been lost. Whether tourism will reappear in the same forms as pre-COVID, or, as some have argued (see for example Special Issue *Tourism Geographies* 2020, 22 (3)), in a greener and more sustainable form remains to be seen. Much will depend on how, when, and in what contexts restrictions are removed and what the tourist image of various destinations is, but it is surely beyond doubt that the pursuit of hope and happiness will resume rapidly and as fully as possible. It is equally surely an obligation for academics studying tourism to pay more attention to the positive side of tourism, not through rose-tinted glasses, not ignoring the warts and all that attract so much negative attention, but to explain why and how tourism can be a means for people to achieve health, happiness and a more positive, hopeful outlook on life.

References

Butcher, J. (2020) The war against tourism. *Spiked*, 4 May. See www.spiked-online.com/2020/05/04/the-war-on-tourism (accessed May 2021).

Butler, R.W. (2021) COVID-19 impacts on the changed and changing nature of the tourism journey. In P. Callot (ed.) *Tourism Post Covid-19: Coping, Negotiating, Leading Change* (pp. 1–16). Montauban: Forestié.

Clawson, M. and Knetsch, J. (1966) *Economics of Outdoor Recreation*. Baltimore: Johns Hopkins Press.

Coppock, J.T. (1977) *Second Homes: Curse or Blessing?* Oxford: Pergamon.

Dann, G.M.S. (1981) Tourist motivation: An appraisal. *Annals of Tourism Research* 8 (2), 187–219.

Dwyer, L. (2022) Tourism development and sustainable well-being: A Beyond GDP perspective. *Journal of Sustainable Tourism* 18. https://doi.org/10.1080/09669582.2020.1825457.
Gartner, W.C. (1994) Image formation process. *Journal of Travel and Tourism Marketing* 2 (2–3), 191–216.
Gretzel, U. (2019) The role of social media in creating and addressing overtourism. In R. Dodds and R.W. Butler (eds) *Overtourism: Issues, Realities and Solutions* (pp. 62–75). Berlin: De Gruyter.
Hall, C.M. and Müller, D.K. (eds) (2018) *The Routledge Handbook of Second Home Tourism and Mobilities*. Abingdon: Routledge.
Lohman, M. and De Bloom, J. (2021) Tourists' benefits from tourism: What we should consider when reshaping tourism. In P. Callot (ed.) *Tourism Post Covid-19: Coping, Negotiating, Leading Change* (pp. 17–30). Montauban: Forestié.
Mihalic, T. (2020) Conceptualising overtourism: A sustainability approach. *Annals of Tourism Research* 84, 103025. https://doi.org/10.1016/j.annals.2020.103025.
Stevenson, R.L. (1902/1878) *El Dorado in Virginius Puerisque*. London: Chatto & Windus
Stevenson, R.L (2009/1877) *An Apology for Idlers*. Penguin: London
Visentin, F. and Bertocchi, D. (2019) Venice: An analysis of tourism excesses in an overtourism icon. In C. Milano, J.M. Cheer and M. Novelli (eds) *Overtourism: Excesses, Discontents and Measures in Travel and Tourism* (pp. 19–38). Wallingford: CABI.
Wheeller, B. (2006) The king is dead. Long live the product: Elvis, authenticity, sustainability and the product life cycle. In R.W. Butler (ed.) *The Tourism Area Life Cycle, Volume 1* (pp. 339–348). Clevedon: Channel View Publications.
Wolfe, R.I. (1977) Summer cottages in Ontario: Purpose-built for an inessential purpose. In T. Coppock (ed) *Second Homes: Curse or Blessing?* (pp. 17–34). London: Pergamon.

Index

access, 24, 95, 112, 162, 171, 186, 195, 204
activity, 5
age, 19, 29, 41, 68, 69, 71, 93, 129, 137, 144, 170, 215, 220, 231, 235
anticipation, 42, 49, 107, 136, 213, 217, 227, 229, 231, 232
attitude, 19, 26, 37, 89, 120, 153, 167, 174

belief, 35, 155, 157, 160, 221, 235
beliefs, 35, 45, 46, 156, 212
benefit, 42, 46, 51, 53, 55, 82, 81, 86, 91, 96, 100, 111, 141, 156, 161, 169, 181, 193, 208
benefits, 7, 12, 13, 21, 40, 41, 44, 49, 55, 57, 58, 60, 62, 83, 88, 86, 91, 95, 100, 104, 107, 113, 116, 117, 135, 137, 141, 143, 145, 148, 153, 171, 172, 173, 174, 175, 177, 181, 182, 183, 198, 227, 229, 236, 238
biases, 232
Buddhism, 34, 35, 41, 42, 44, 47, 49

camp, 161, 162, 164, 211, 217
Canoe, 203, 204, 206, 210, 211, 213, 214, 216, 217, 218, 220, 221, 222, 223, 227
canoeing, 203, 204, 206, 214, 221, 222, 223, 225, 227
celebrity, 42
children, 38, 51, 80, 106, 10, 11, 111, 112, 113, 114, 118, 144, 179, 218, 231, 233
China, 25, 94, 118, 181, 200, 203

destination, 12, 13, 14, 19, 24, 44, 45, 46, 48, 65, 67, 71, 78, 80, 81, 82, 83, 84, 86, 88, 89, 90, 91, 94, 95, 96, 99, 100, 107, 114, 137, 140, 173, 175, 179, 182, 183, 184, 210, 216, 227, 229, 231, 235
destinations, 8, 9, 16, 17, 21, 22, 23, 25, 32, 69, 75, 80, 83, 91, 94, 95, 97, 113, 114, 155, 156, 175, 177, 179, 182, 183, 188, 201, 204, 227, 229, 231, 232, 236
diet, 147, 211

effort, 39, 76, 156, 175, 220, 235
ego, 88, 89, 99
enjoyment, 9, 86, 114, 120, 129, 137, 182, 208, 220, 229, 231, 233, 235
environment, 27, 48, 53, 58, 63, 75, 89, 90, 140, 147, 172, 174, 175, 177, 180, 181, 190, 210, 214
exercise, 17, 23, 29, 40, 86, 159, 169, 190

facilities, 95, 97
facility, 159
fitness, 52, 141, 145, 147, 148, 177, 211, 220
food, 18, 21, 32, 42, 48, 86, 88, 89, 90, 91, 93, 94, 95, 96, 97, 98, 99, 100, 101, 102, 103, 113, 116, 147, 186, 204, 213, 214, 215, 216, 218, 222, 231
foods, 114
fulfillment, 233

happiness, 5, 6, 7, 8, 11, 13, 14, 28, 32, 35, 51, 53, 54, 55, 57, 58, 60, 62, 64, 63, 65, 67, 83, 81, 83, 86, 88, 90, 91, 94, 97, 98, 99, 100, 101, 109, 111, 114, 118, 120, 126, 127, 128, 129, 135, 136, 137, 140, 143, 148, 150, 152, 154, 155, 153, 159, 167, 169, 171, 172, 179, 197, 203, 209, 213, 221, 225, 231, 235, 236
health, 3, 9, 11, 12, 22, 44, 91, 101, 104, 107, 116, 117, 141, 142, 143, 144, 145, 146, 147, 150, 152, 172, 171, 172, 173, 175, 177, 179, 180, 183, 181, 182, 183, 209, 213, 217, 235, 236

heritage, 42, 69, 84, 88, 91, 97, 101, 114, 155, 156, 157, 159, 160, 164, 169, 170, 171
Hope, 49, 55, 62, 61, 63, 65, 67, 69, 71, 74, 76, 78, 82, 81, 84, 88, 185, 187, 189, 191, 193, 196, 198, 200, 201, 202, 203, 229, 232, 233, 236

illness, 106, 107
imagination, 153, 164
India, 32, 88, 91, 101, 103, 202

journey, 32, 119, 145, 146, 147, 148, 191, 204, 220, 222, 225, 236
joy, 9, 97, 109, 126, 128, 129, 141, 145, 147, 148, 209, 213, 227, 233, 235, 236

lifestyle, 18, 83, 89, 98, 99, 120, 141, 147, 172
lifestyles, 89, 172
link, 8, 35, 39, 57, 78, 81, 99, 129, 166, 173, 175, 179, 203, 210, 214

Management, 12, 13, 14, 48, 49, 61, 63, 83, 84, 86, 88, 101, 102, 103, 117, 118, 142, 150, 152, 154, 171, 182, 181, 183, 184, 200, 222, 227
meaning, 4, 5, 6, 13, 14, 16, 17, 20, 21, 23, 24, 25, 26, 28, 29, 51, 68, 107, 109, 128, 129, 130, 140, 143, 144, 145, 157, 159, 164, 171, 173, 174, 183, 212, 216, 217, 231, 232, 231
media, 48, 81, 152, 232, 231, 238
Meditation, 34, 40, 47, 48, 142, 147, 152
memories, 8, 23, 74, 75, 78, 83, 96, 104, 111, 112, 113, 120, 153, 155, 156, 157, 159, 167, 222, 229
memory, 18, 37, 44, 114, 116, 153, 157, 170, 171
mental, 18, 89, 100, 104, 107, 109, 113, 116, 143, 174, 175, 177, 181, 211, 217, 220, 233
mind, 4, 17, 29, 33, 36, 38, 41, 47, 54, 68, 86, 97, 107, 129, 175, 204, 203, 208, 210, 211, 216, 217, 218, 222, 227, 229, 233
mindfulness, 32, 33, 34, 35, 36, 37, 38, 39, 40, 41, 42, 43, 44, 45, 46, 47, 48, 49, 146, 173, 203, 206, 219, 221, 225, 229, 231

movement, 123, 166, 191, 195, 214, 236

natural, 10, 17, 29, 42, 58, 89, 100, 143, 147, 148, 157, 172, 174, 175, 177, 179, 180, 181, 182, 184, 189, 208, 210, 213, 214, 216, 218, 222, 221, 223, 232
nature, 5, 11, 12, 30, 49, 51, 58, 60, 62, 64, 71, 76, 81, 91, 95, 97, 98, 11, 113, 117, 118, 128, 129, 136, 140, 141, 146, 147, 148, 150, 152, 174, 175, 177, 180, 181, 185, 195, 197, 203, 204, 206, 208, 209, 213, 217, 219, 222, 221, 222, 223, 227, 236

outcome, 6, 10, 52, 57, 60, 71, 80, 86, 162, 164, 167, 173, 210
outcomes, 4, 9, 19, 33, 34, 35, 37, 38, 39, 40, 41, 44, 51, 55, 60, 61, 78, 106, 107, 120, 137, 141, 145, 155, 157, 159, 166, 167, 169, 171, 173, 175, 179, 181, 197, 209

path, 10, 16, 91, 95, 145, 146, 148
pathway, 29, 34, 35, 39, 42, 68, 83, 81, 155, 153, 155, 157, 165, 167, 169, 170
peace, 144, 147, 172, 185, 186, 187, 188, 190, 191, 193, 196, 198, 200, 201, 202, 204, 203, 204, 206, 208, 212, 213, 211, 213, 216, 218, 219, 220, 221, 222, 223, 227, 229, 233
perception, 19, 20, 38, 64, 175, 183, 184
perceptions, 10, 45, 55, 64, 63, 84, 101, 107, 186
physical, 3, 14, 20, 23, 38, 40, 69, 71, 75, 86, 89, 90, 91, 95, 104, 109, 113, 129, 141, 143, 145, 162, 172, 171, 173, 175, 177, 180, 186, 204, 210, 213, 211, 217, 220, 233
pleasure, 5, 7, 10, 12, 13, 23, 28, 30, 58, 89, 90, 99, 101, 107, 109, 117, 123, 135, 136, 137, 141, 148, 150, 155, 156, 160, 169, 171, 183, 210, 213, 214, 215, 222, 227, 229, 231, 232, 231, 235, 236
positive psychology, 3, 4, 5, 6, 7, 8, 9, 10, 11, 14, 58, 84, 86, 88, 91, 96, 100, 101, 104, 107, 109, 114, 117, 118, 146, 155, 167, 170, 171, 172, 177, 180, 183, 181, 183, 233
privacy, 8
private, 11, 126, 162, 223

problem, 18, 19, 30, 38, 51, 68, 231
problems, 29, 34, 38, 40, 45, 71, 91, 120, 162, 198, 205, 218, 227, 232
process, 10, 19, 27, 47, 51, 54, 71, 90, 92, 101, 140, 141, 143, 145, 146, 147, 148, 153, 159, 167, 175, 184, 188, 193, 195, 196, 197, 198, 211, 212, 238
processes, 9, 18, 19, 34, 38, 39, 64, 75, 90, 100, 147, 183
public, 69, 71, 103, 120, 121, 200, 223

Recreation, 12, 15, 30, 48, 49, 61, 63, 64, 102, 103, 117, 118, 150, 152, 206, 208, 222, 223, 224, 225, 227, 235, 236
recuperate, 141, 144
Recuperation, 142
relax, 109, 112, 141, 146
relaxation, 5, 40, 111, 136, 138, 141, 148, 175, 183, 227, 231, 233, 235
resilience, 80, 83, 84, 109, 181, 210, 220
Resilient, 184
Resort, 25, 30
resorts, 93
resource, 30, 76, 103, 130, 184, 206, 210, 223, 227
rest, 10, 120, 127, 141, 144, 146, 147, 148, 175, 208, 227
retreat, 13, 49, 135, 141, 142, 143, 144, 145, 147, 148, 150, 152, 233
rural, 42, 48, 67, 68, 69, 86, 88, 89, 90, 91, 94, 97, 99, 100, 101, 103, 143, 161, 172, 231

self, 3, 6, 11, 15, 22, 23, 25, 26, 27, 32, 35, 37, 40, 41, 45, 51, 52, 53, 54, 57, 60, 65, 68, 69, 74, 81, 86, 88, 90, 99, 100, 106, 109, 121, 135, 136, 137, 138, 140, 141, 142, 143, 144, 146, 147, 148, 150, 152, 154, 156, 160, 161, 172, 174, 177, 183, 193, 196, 204, 210, 212, 213, 219, 220, 222, 221, 229, 231, 233, 235
sensory, 18, 78, 86, 91, 103, 123, 157
services, 9, 118, 162, 164, 171, 180, 182, 196
solitude, 8, 89, 97, 204, 213, 217
spa, 14, 171, 231
sustainability, 16, 21, 32, 34, 40, 44, 45, 46, 47, 48, 49, 130, 180, 181, 185, 222, 225, 238
sustainable, 16, 21, 22, 45, 46, 47, 53, 58, 62, 63, 86, 99, 103, 167, 183, 211, 218, 223, 229, 236, 238

togetherness, 8
trail, 208, 210, 215
trails, 171, 210
transport, 23, 66
travel, 4, 5, 9, 10, 11, 12, 16, 22, 23, 24, 28, 30, 32, 42, 46, 49, 53, 58, 66, 80, 83, 89, 93, 94, 97, 104, 106, 107, 108, 112, 113, 114, 116, 117, 118, 120, 130, 141, 147, 150, 152, 154, 155, 153, 171, 172, 179, 182, 183, 184, 185, 188, 189, 217, 221, 223, 229, 232, 233, 235, 236

volunteer tourism, 49, 51, 53, 54, 55, 57, 58, 60, 61, 62, 63, 64, 138, 172, 231

Walk, 206, 222
Wellbeing, 146, 147, 150
Wellness, 15, 31, 61, 144, 146, 147, 152, 154, 180, 181
wilderness, 177, 206, 208, 211, 212, 213, 217, 220, 222, 223, 225
Wildlife, 143, 183

For Product Safety Concerns and Information please contact our EU Authorised Representative:

Easy Access System Europe

Mustamäe tee 50

10621 Tallinn

Estonia

gpsr.requests@easproject.com